ELEPHANT

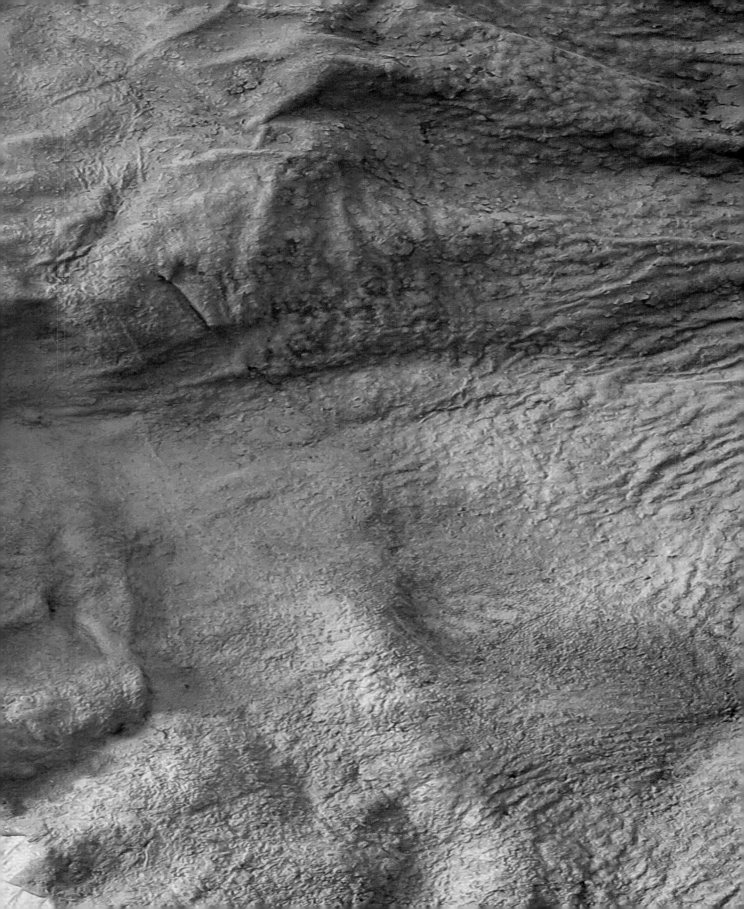

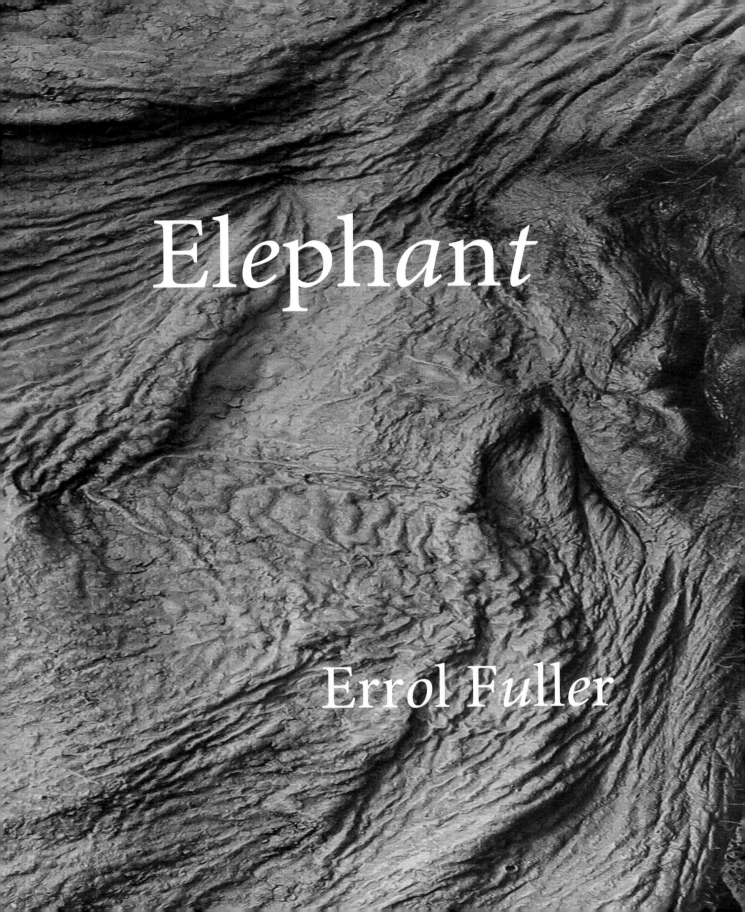

Elephant

Errol Fuller

Copyright © 2019 by Princeton University Press

Published by Princeton University Press
41 William Street, Princeton, New Jersey 08540
6 Oxford Street, Woodstock, Oxfordshire OX20 1TR

press.princeton.edu

All rights reserved

Library of Congress Control Number 2018965184
ISBN 978-0-691-19132-4

British Library Cataloging-in-Publication Data is available

Editorial: Robert Kirk and Kristin Zodrow
Production Editorial: Mark Bellis
Design: Errol Fuller
Publicity: Sara Henning-Stout and Julia Hall
Copy Editor: Laurel Anderton
Jacket Credit: David Chancellor

This book has been composed in Minion Pro

The publisher would like to acknowledge the author of this volume for providing the print-ready files from which this book was printed.

Printed on acid-free paper

Printed in China

10 9 8 7 6 5 4 3 2 1

For Dr. Prasert Prasarttong-Osoth,
a man who loves elephants

Acknowledgements

As with most books there are a number of people whose help has been invaluable. The first is Dr. Prasert Prasarttong-Osoth without whose great enthusiasm and assistance this book would not have happened. Similarly, I have to thank Piriya Vachajitpan.

This particular book is as much about images as it is about words, and several photographers have kindly allowed use of some of their wonderful images. Without their kindness the book would not exist. In alphabetical order, they are: Annick at Aldo Workshop, Tracey Barclay, Peter Beard, Thomas Breuer, David Chancellor, Glyn Clarkson, Granville Davies, Tim Flack, John Hodges, John Metcalf, Pat Morris, Bury Peerless, Rattapol Sirijirasuk.

Three artists also need to be thanked for allowing use of their paintings – Raymond Ching, Prateep Kochabua, and Walton Ford. Others who have kindly helped are: Jill Austin, Hilary Knight, Lyulph Lubbock, Diana Maclean, Peter Petrou, and Thula Thula Game Reserve.

Finally, I have to thank my good friend Irene Palmer, who has not only contributed photographs, but also helped with many invaluable snippets of information.

If I have forgotten anyone – I am very sorry.

Dust wrapper. Elephants in northern Kenya. Photo by David Chancellor.
Endpapers. An elephant's eye. Photo by John Hodges.
Half title. The face of an Asiatic elephant. Photo by Tracey Barclay.
Title page. Elephant skin. Photo by Annick at Aldo Workshop.
Previous page. An Asiatic elephant at bay. Photo by John Hodges.
Facing page. Marsh Elephant. Photo by Tim Flack.

This beast passes all others in wit and wisdom
                              Aristotle (circa 350 BC)

Facing page. A night-time procession in Sri Lanka. Photo by Bury Peerless.
Two following spreads. Elephants at Etosha National Park, Namibia. Photos by Irene Palmer.
Pages 14 and 15. A pregnant Indian elephant deep in the jungle. Photo by John Hodges.
Pages 16 and 17. An elephant having fun. Photo by Pat Morris.
Two spreads following page 17.
   A monolithic seventh-century Indian elephant sculpture at Mahabalipuram. Photo by Bury Peerless.
   An Asian elephant helping with logging. Photo by Bury Peerless.

This page and facing page. Antique sculpture of an elephant at Sukhothai, the ancient capital of Thailand.

CONTENTS

Elephant — 26

African Elephants — 44

Asian Elephants — 78

In the Beginning – Evolution, Mammoths, Mastodons — 108

Elephant Curiosities — 138

The Elephant in Art, Literature, and Popular Culture — 192

Conservation — 248

The Elephant is comes nearest

Pages 26 and 27. Indian Elephant by Raymond Ching. Oil on panel, 10 and a half inches x 15 inches (27cm x 37cm).

the beast that to man in intelligence

Pliny the Elder (circa AD 77)

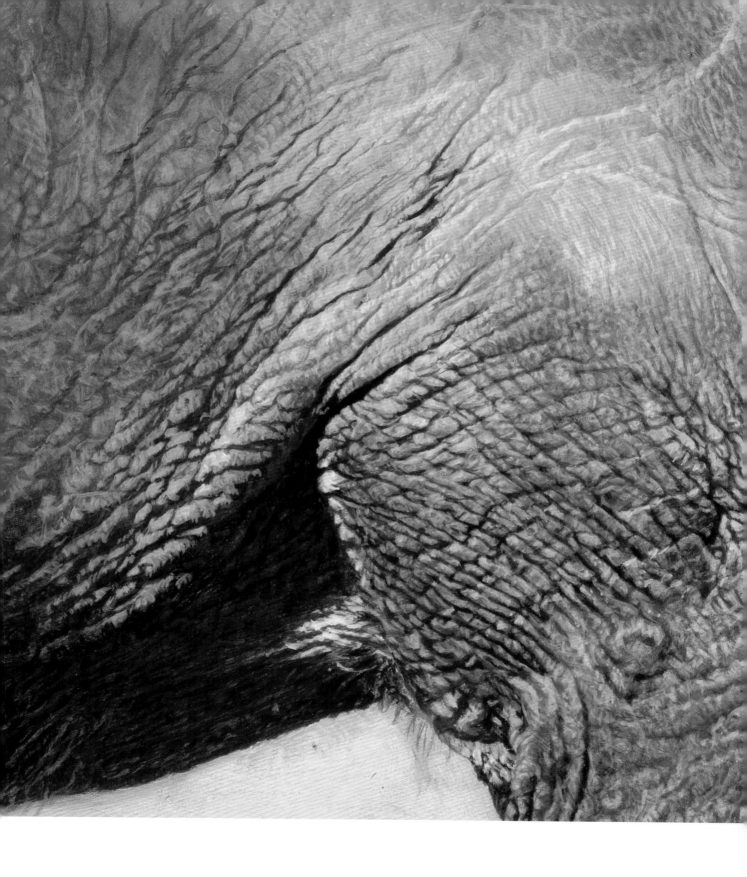

Elephant

Of all creatures living in the world today, elephants are among the most easily recognizable. Visually they are instantly familiar, yet despite this comforting familiarity they confront us with so many mysteries and contradictions. The saddest of all these contradictions is certainly an obvious one: popularity does not equate with safety. It affords elephants only minimal protection from the depredations of

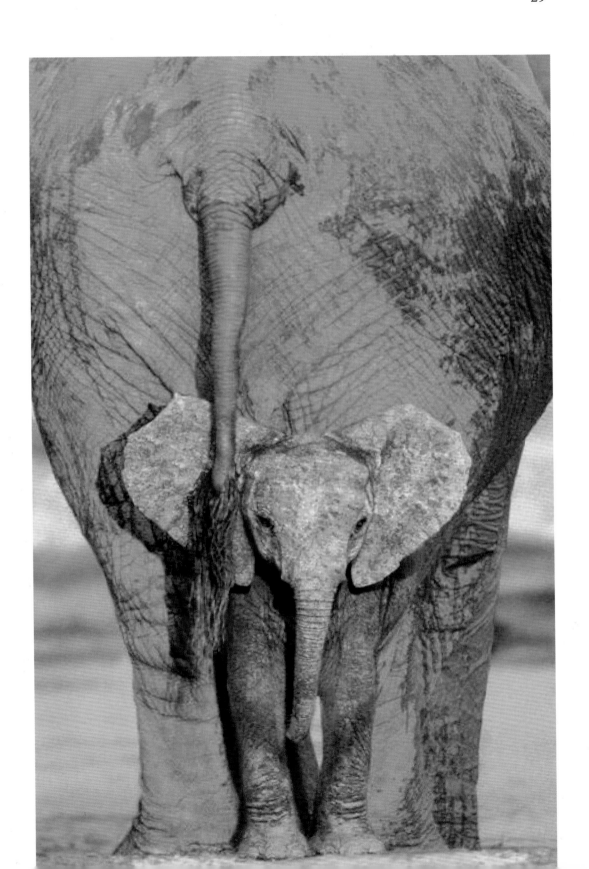

humans – for they are among the most viciously persecuted of all animals.

Other contradictions, although not so disturbing, are just as evident. While even very young children can identify the image or shape of an elephant – and respond to it in the most favourable ways – there remain so many aspects of elephant activity that even the most

Page 28. The face of an Indian elephant. Photo by John Hodges.
Page 29. A baby elephant with 'butterfly' ears sheltering beneath its mother's legs.
Facing page. Elephant and holy man. Photo by Bury Peerless.
Above. A white elephant on the flag of Siam from 1855 to 1916.

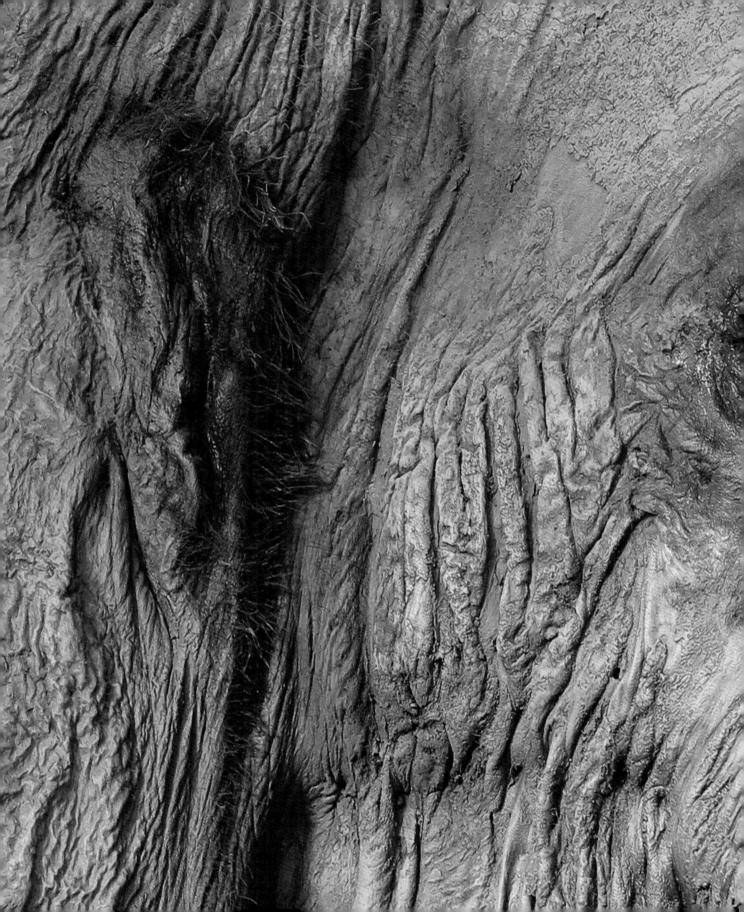

sophisticated of animal behaviourists struggle to understand. They display, for instance, traits that appear remarkably human-like, often showing emotions that we can all relate to: loyalty, playfulness, long memories, secrecy, the occasional desire for revenge, the ability to grieve, and a seeming self-awareness. They can even be taught to perform physically demanding tasks of considerable use to humans. Add to this, their development of a social system that in some respects seems to closely echo our own.

Even their so familiar – albeit prehistoric – appearance holds inherent fascination. The trunk that can be used almost like a hand, the massive curving tusks, the thickly folded yet curiously baggy skin, the enormous flappy ears, the long pillar-like legs that seem to stand almost vertically beneath the mass of the body – all these features set elephants apart from other animals.

Pages 32 and 33. Skin. Photo by Annick at Aldo Workshop.
Facing page. Trunk and face. Photo by John Hodges.
This page. Legs. Photo by Irene Palmer.
Following two pages. The Elephant's Eye. Photo by John Hodges.

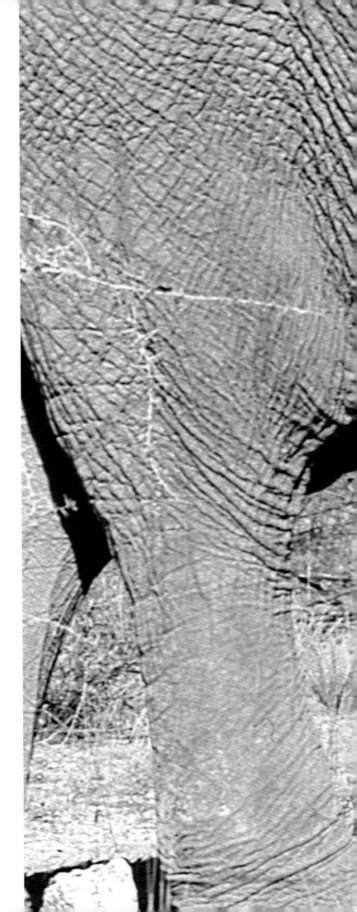

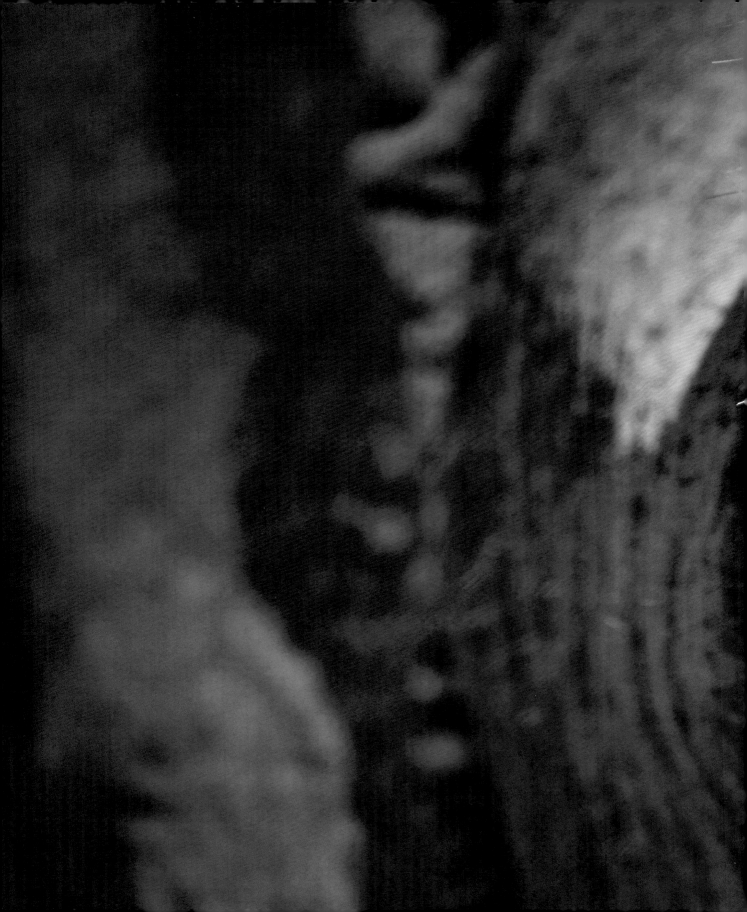

This page. The Elephant's Ear. Photo by Granville Davies.

Facing page. The Elephant's Foot. Photo by Pat Morris.

And then there is the dark side. Humans have relentlessly persecuted these giant creatures for centuries. From our perspective some of this persecution has been inevitable; elephant herds can be enormously destructive to vegetation, and to the land itself. As humans became increasingly reliant on agriculture, the interests of the two species came more and more into conflict, and when people collide with animals there is usually only one ultimate winner.

Less easy to comprehend is the human obsession with ivory, an

obsession that has led to the wholesale destruction of elephants on an almost industrial scale. Regrettably, such destruction is by no means at an end.

Leaving this horror aside, the curiosities that surround elephants are many and various. Some are comparatively insignificant yet still intriguing: facts like the elephant habit of napping while standing up but lying down to sleep properly. Others are perhaps more profound. What, for instance, is to be made of the idea of elephant graveyards? And just as tantalizing is the subject of elephant painting. Unlike any other animal, these creatures can paint pictures (using their trunks) in which the subjects – flowers, trees, houses – are immediately recognizable. What does this surprising ability mean?

Then, of course, there is the matter that gives elephants their true star quality. They are, by a considerable margin, the largest of all living terrestrial creatures.

Page 40. Tusks.
Page 41. The Elephant's Head. Photo by Irene Palmer.
Facing page. The feet of an Indian elephant. Photo by John Hodges.

African Elephants

African Elephants

The charge of an African bull elephant can happen with little warning, or even none at all. The colossal animal can turn with devastating speed and move towards whatever has aroused his ire with frightening purpose. His head suddenly rears upwards, raising the tusks that no longer seem like benign ornaments; now they assume the role of awesome weaponry. The eyes start from the head like burning coals as they stare forward and focus on their target. The ears slap loudly against his body, then perform a peculiar dance of their own, and in fury he may let out a fearsome trumpeted bellow or, even worse, he may move forward with absolutely silent, but deadly, intent.

For the African bush elephant (*Loxodonta africana*) is the world's largest and most imposing terrestrial creature, a creature of real flesh, blood, and passion; when it charges it no longer conforms to the imaginings so many people have – of a large and lovable cuddly toy that has magically come to life! Now it is a creature with terrifying physical power, even if it chooses to use that power comparatively rarely.

Pages 44 and 45. An elephant at Etosha, Namibia. Photo by Irene Palmer.
Facing page. A large bull elephant charging. There are perhaps two reasons for the rather blurry nature of the image – the powerful movement of the animal itself, and the circumstance in which the photographer found himself. A truly evocative photo by David Chancellor.
Following two pages. An African herd. Photo by Granville Davies.

50   AFRICAN ELEPHANTS

It is commonly said that of all African species dangerous to humans, the most feared are the hippopotamus or the Cape buffalo. If you talk to local people, however, most of them will tell another tale. The elephant is the animal they will accuse, and particularly individual animals that are alone.

Above. An elephant confronted by two rhinos at Thula Thula Game Reserve, South Africa. This remarkable photo was taken by a trainee ranger at the reserve named Jenni Smith.

The statistics are impressive. A very large male can stand 13 feet (almost 4 metres) at the shoulder and, taking into account the tusks and tail, its length can be measured at up to 30 feet (a little over 9 metres). The tusks alone have been known to measure 10 feet (3 metres) from base to tip and have weighed in at 200 lbs (90 kg);

Above. A bull elephant emerges from the water. Photo by David Chancellor.

the total weight of the animal can reach 15,000 lbs (700 kg). In terms of longevity, elephants can live for 70 years, and the gestation period is around 22 months. Essentially, elephants have three main requirements – water (and plenty of it), salt (they need it to maintain health), and masses of edible vegetable matter. Each animal might eat as much as 650 lbs (300 kg) of food a day.

Leaving aside the other statistics, these vast food requirements form part of the problem existing between people and elephants for, like humans, these creatures are capable of radically altering their environment. The destruction of plant life caused by elephant herds in the constant search for enough food to maintain their bulk is easy to imagine, and not only might this interfere with agricultural projects, it can easily result in general habitat alteration.

Comparatively recently Africa was one vast wild continent over which elephants could roam at will, and with such an immense area available the balance between animals, vegetation, and the land itself was sustainable. Now things are different.

Human population growth combined with rapid technological advance creates and facilitates ever-increasing needs that are in almost total conflict with the maintenance of the vast undisturbed wild regions that once characterized much of Africa. Conservationists and governments will claim that large areas have been given over to

Previous two pages. The natural order – vultures and marabou storks feeding on a carcase. Photo by Irene Palmer.
Facing page. The awful result of unsustainable land use. Photo by Peter Beard.
Pages 56 and 57. Elephants at Etosha, Namibia. Photo by Irene Palmer.

form wildlife reserves – and this is perfectly true. However, this is not necessarily the wonderful solution that many believe it to be. Though the areas may be large they are finite, and the damage that elephants can do in such places can get out of hand, with unforeseen consequences in terms of habitat destruction.

Plant life can be drastically affected, causing radical alteration to the land. This in turn can effect all kinds of changes in patterns of wildlife behaviour – many of them unpredictable – that in turn can lead to species disappearing entirely from places where they were once abundant.

At the start of his influential book *Song of the Dodo* (1996) David Quammen outlined the problem in an illuminating and disturbingly revealing way:

> *Let's start by imagining a fine Persian carpet and a hunting knife. The carpet is twelve feet by eighteen, say. That gives us 116 square feet of continuous woven material… We set about cutting the carpet into 36 equal pieces, each one a rectangle, two feet by three… When we're finished cutting, we measure the individual pieces, total them up – and find that, lo, there's still nearly 116 square feet of recognizable carpetlike stuff. But what does it amount to? Have we got 36 nice Persian throw rugs? No. All we're left with is three dozen fragments, each one worthless and commencing to come apart. Now take the same logic outdoors and it begins to explain why the tiger has disappeared from the island of Bali. It casts light on the fact that the red fox is missing from Bryce Canyon National Park… An ecosystem is a tapestry of species and relationships. Chop away a section, isolate that section, and there arises the problem of unravelling.*

So, the alteration elephants can cause when restricted to areas that are essentially limited in extent can be catastrophic, and this can be to the detriment of many other species – not just to humans.

Above. A lone elephant in barren country. Photo by Granville Davies.

The problem between human and elephant is, of course, not just competition for habitat; it also revolves around the human greed for ivory. Because of this greed, elephants have suffered for centuries, and despite the efforts of conservationists it is a lust that still continues. While it may be true to say that hunting of elephants for tusks has less effect on overall population size than the general decrease in available land, the horrible individual suffering that it causes is a monstrous blot on humanity.

There are many respects in which the social life of the African elephant resembles that of humans. It can be entirely misleading to equate animal behaviour – motives, planning, interaction, gestures, and so on – with our own, but in the case of elephants it is a temptation difficult to resist, and there is every likelihood that such comparisons give a clearer and deeper insight into the lives of these creatures. However, there is one obvious way in which their behaviour differs radically from the system adopted by most human societies (although certainly not all!); male and female elephants spend their lives in groups that keep themselves essentially separate from one another.

Females live together with their young in closely knit family groups consisting of a varying number of individuals all led by a matriarch. This matriarch appears to make all the decisions for the well-being of the community – where to go, when to do it, how to interact with rival groups and so on. She will usually be the oldest family member and she will stay in control until her death, or until she becomes too decrepit to offer suitable leadership. A ranking hierarchy will have developed and usually it will be the next oldest member of the troupe who inherits the vacant crown.

Facing page. A photograph shot during the 1890s showing two huge tusks taken from an elephant killed on the slopes of Mount Kilimanjaro.
Following two pages. Elephants moving silently through the land.

The group will often cooperate as a unit in all manner of ways for the good of all or for the good of an individual. If, for instance, a baby falls into difficulty at a waterhole, the adults may all pitch in to help get the youngster to safety. Then, once it is out of trouble, the baby is likely to run to be soothed by its mother, or it may turn to an older sister or an aunt for bodily comfort and reassurance.

Babies are in fact born into a comparatively secure and safe situation considering the dangers that are all around. Within minutes of its birth, an elephant can stand and scamper alongside its mother, although it can make a rather strange sight as control over its trunk is limited and it may just flap about. It is a long time before the young animal gains full control over its extraordinary appendage.

Even so, it can bellow with a surprising level of volume, an ability that is obviously very necessary, for at this period of its life – despite the relative security provided by the family unit – it is vulnerable, and the need to issue out a warning or make a cry for help may be a fairly regular one. Some have suggested that milk tusks develop, and that these eventually drop out to be replaced by the real thing. But this notion is something of a controversial matter and is not universally accepted.

Sometimes a family group will come together with other groups to create a loose association that has come to be termed a bond group. These seem to be comparatively temporary arrangements, but when they meet up after an unspecified interval, the coming together can be like a human family reunion.

The youngsters will run about playfully, moving beneath and between the adults, and will play right beside them just as human children will often irritate adults by not moving off to a more respectful and peace-inducing distance. Meanwhile the mature females will engage in ritual greetings – touching, nodding and swaying, and even appearing to indulge in chatter and gossip. When necessary, these large assemblages will defend their perceived territories against other bond groups.

Facing page. Two elephants embracing. Photo taken at the Thula Thula Reserve by guest Chelsea Ropes.

While the females will grow up and stay in the family unit for their whole lives, the young males stay for just a few years, sometimes eight or more, sometimes considerably longer. But at some stage these young bulls begin to test their female relatives in irritating ways.

Eventually the elders tire of the constant discord and encourage, or even force, their young relatives to leave. When this finally happens the rejected animal wanders off.

In some cases the bull will remain solitary, but alternatively he may choose to seek out company. If he chooses the latter path he will go in search of a group of bulls, hoping to find a troupe that will allow him to join their gang.

Like the female families, this group will be strongly hierarchical, with a dominant leader and beneath him a pecking order of lesser bulls all vying for position and respect. Joining such a group may not be easy.

Various cautious approaches may be necessary. Initial rejection is likely and persistence may be needed. Approval of the lead bull may be gained by a variety of means. These may include first gaining the support of lesser dignitaries, or by small acts of general respect and submission. Even such a human trait as personal charm may have its place!

Facing page. A small herd, looking beleaguered. Photo by Peter Beard.
Pages 68 and 69. A herd of elephants in the Ewaso Nyiro River of northern Kenya. This river (the name translates as 'muddy waters' or 'muddy river') system is of vital importance to the wildlife of the area. Photo by David Chancellor.

The complexity of relationships in groups of bulls is steeped in ritual, and acts of friendship or conciliation are well developed. One such act that indicates affection or submission is for a bull to dip the end of its trunk into the mouth of another, a performance that almost corresponds to a human handshake.

Although the individuals in the group may look remarkably similar to the untutored eye – some bigger, some smaller – there are all sorts of subtle differences; some will have scars, some nicks out of their ears, and some will have differently shaped tusks.

Assuming a young bull is accepted into a group he then stays with his new friends and conforms to life in a highly structured society where each individual knows his place in the pecking order. Naturally, the structure and the pecking order may change with time (and membership of the group may itself be quite fluid), but the status quo is usually maintained, and under normal circumstances this is a life that is led, essentially, without female company.

Bearing in mind this arrangement an obvious question rears its head. How do elephants breed? The answer is simple: male elephants are periodically overwhelmed with a massive rush of testosterone, and this rush has been given the name *musth*. It will cause an animal to go in search of a female receptive to his needs or, if he is a member of a group, to wander off and do the same.

Despite the extremely hierarchical nature of male society, when a bull comes into *musth* he may forget his place. His behaviour becomes unpredictable and aggressive and at his approach higher-ranking bulls may choose to give way. For the time being they may overlook his disregard for the social order and be prepared to defer to their social inferior. Under the influence of *musth* the affected animal exudes a distracted yet powerful personal aura and gives off clear signals that reveal the state he is in. As he moves he dribbles urine, his ears wave as he curls his trunk and he indulges in exaggerated displays of aggression, and eventually he moves off in search of a female who is herself in a state ready to receive his advances.

This system seems to confer an advantage in terms of keeping the gene pool varied, for the bulls come into *musth* at different times. So, a young bull may arrive at the state of *musth* when those above him in the social order (including the leader) are in a more normal condition and have no sexual interest. As a result the individual in *musth* may have only limited competition in his search for a female, a competition that he would almost certainly lose if his social superiors were taking part. The leader would assert his dominance, and if this were the case only this leader would get to breed. Matters are not quite as simple as this sounds, however, for although the *musth* periods may often be staggered, there are often overlaps, and then the competition can be intense. Also, there is some evidence to suggest that females may be more sexually receptive at times of peak food production and that the larger males somehow synchronize their *musth* period with this. So, as with many things in life, the pattern is not as firmly regulated as it could be!

All of this is a simplified summary of what goes on in the lives of these elephants but, of course, just as with the mating behaviour, everyday life does not always conform to standardized schemes. And although it may be a description of the basic way of life for many, it certainly doesn't apply to all.

Some live in an environment quite different from the open country generally associated with elephants. These less familiar animals live deep in the forest and have been given the appropriate name of 'forest elephants'.

Even though they may constitute as much as a third of the total population of all the elephants living in Africa, they remain something of a mystery. Partly, this is because they are much more difficult to see and track than those elephants living in more open country. They inhabit the forests of central equatorial Africa and, despite their size, in such areas their presence is not always easy to detect.

For many years they were regarded as belonging to the same species as the familiar African bush elephant (*Loxodonta africana*), but in recent years they have been recognized by most zoologists as being distinct enough to warrant separation.

Naturalists have argued for decades over whether or not elephant populations should be divided into separate species, and now current thinking among most elephant biologists is that the forest animals should be regarded as quite separate from those that inhabit more open country. They are now usually given full species rank and their scientific name is *Loxodonta cyclotis*.

Although so similar in general appearance there are significant physical differences between the two species. Forest elephants are darker in colour and smaller in size – a large male might stand 8 feet (2.5 metres) at the shoulder, compared to the 13 feet

Previous two pages. Elephant eyelashes. Photo by Tim Flack.
Facing page. A forest elephant with its young in the Mboli River, Nouabale-Ndoki National Park, Congo. Photo by Thomas Breue

(almost 4 metres) that a large bush elephant might reach. The tusks are likely to be harder, yellower in colour, stronger, and straighter. In proportion to their size they may also be longer, sometimes almost reaching to the ground. These tusk features are thought to be helpful in the animal's need to push through patches of dense vegetation.

Social arrangements are not entirely dissimilar to those of elephants living in more open areas, although the adult males seem to be more likely to be solitary. Females and young animals move about in small groups usually consisting of no more than eight individuals.

Although they are strictly vegetarian, the diet is very varied due to their forest habitat, and it is thought that this makes these elephants very important distributors and dispersers of seeds during their constant travels, a factor that may have very important, and beneficial, environmental consequences.

Some particularly small individuals occur in the heart of forests of the Congo Basin, and these animals are often referred to as 'pygmy elephants.' These comparatively miniature creatures are considered by some researchers to form an entirely separate species, but the general consensus is that they are merely forest elephants of remarkably small size and that this size has developed due to environmental circumstance.

On the subject of the splitting of some elephant groups into separate species, it is worthwhile to mention that there was once a distinct population living in the Atlas Mountains of North Africa – and some researchers consider that this population actually constituted another distinct species. However this may be, there is no doubt that these elephants are now extinct. Curiously, it was these particular animals that were tamed and famously used by Hannibal during the Punic Wars when the Carthaginians attacked the Roman Empire and marched their war elephants across the Alps in 218 BC.

Above. An elephant at Etosha, Namibia. This area is a natural salt flat, and because of the mineral content on the surface it is very attractive to elephants. It lends the animals a slightly ghostly apperance due to their dusting with salt. This pale look is particularly evident in many photos taken there by Irene Palmer.

Asian Elephants

Asian Elephants

The elephants of Asia are superficially similar to those of Africa, the differences being rather subtle. In general size they are smaller than African bush elephants, as are their ears, and many individuals have two pronounced lumps on the forehead; other than this there is visually little to tell them apart, although females do not generally sport tusks.

Those who specialise in systematics have decided that Asian elephants are more closely related to the extinct mammoth than they are to their African cousins.

But as far as humans are concerned the chief difference lies in the fact that Asiatic elephants are tractable in ways in which African elephants seem not to be. This may, however, be due to the cultural and traditional habits of Asian peoples rather than to the nature of the animals themselves, and there is evidence to show that some African elephants have certainly been trained to serve humans. Inspired by the pioneering work of missionaries in Gabon – who had succeeded in capturing and training a young forest elephant – Leopold II, King of the Belgians, sponsored a training programme at the turn of the

Pages 78 and 79. The beginning of a nocturnal procession with dressed elephants and flags. Elephants have allowed themselves to be dressed in ways similar to this for centuries. Whether or not they enjoy the process, is not known. Photo by Bury Peerless.
Facing page. Elephant and man playing together at Jaipur. Photo by Tracey Barclay.
Following two pages. Elephant washing time. Photo by Pat Morris.

nineteenth century, a programme that achieved considerable success. There are certainly other examples.

However this might be, the elephants of India and southeast Asia have for centuries been celebrated for their usefulness and trained to perform a variety of tasks in the service of humans – and so have acquired reputations for being much more tractable beasts than their African counterparts.

Such tasks include lifting and moving heavy items, providing entertainment for humans, acting as transportation for objects and people, serving as awesome weaponry in warfare, and even being trained as executioners.

Pages 84 and 85. Elephant faces. Photo by John Hodges.
This page. A painted elephant at Jaipur, India. Photo by Bury Peerless.
Facing page. Captive elephants and their babies. Photo by Pat Morris.
Pages 88 and 89. A captive elephant hiding (as they sometimes do) in the woods. Photo by John Hodges.

Above. An elephant moving very heavy timber. Photo by Pat Morris.

For all of these purposes many elephants have lived their entire lives in captivity.

This lasting and complex relationship was established between humans and elephants so long ago that its origins are lost in the mists of time, and it is not just a matter of the usefulness of the animals in terms of labour. The relationship also assumes cultural and religious significance.

In terms of control, ancient chronicles exist that show 'maps' of the elephant body, and these diagrams are carefully marked to show many points of sensitivity. They are basically indications of the presence of nerve centres that could be pressed or pulled by mahouts to create a response in an animal.

Although captivity is so common, some degree of the awesome potential and underlying wildness of Asian elephants is revealed in the way in which many of those familiar with the animals categorize them. Owners often consider that elephants fall into one of three basic types.

The first is classified as never being dangerous. The second is classed as only being dangerous under certain conditions (e.g., being surprised by disturbing or loud noises, being distressed by the absence of a regular mahout, an occasional uneasiness in water, etc.), while the third class is considered as being potentially dangerous at any time.

Facing page. Mother and baby. Photo by Pat Morris.
Following two pages. A family of wild elephants. Photo by Rattapol Sirijirasuk.

In the wild, their way of life is not dissimilar to that of their African relatives.

Adult females live in family groups with their young, often travelling in large associations of several families. Each family will have a leader, usually one of its oldest members, and as the males reach adulthood they will usually wander off, either staying solitary or else joining a group of other males.

Their thick skin affords protection in many obvious ways, but it can also be something of a curse as parasites of many kinds may seek shelter in the deep folds and cause considerable discomfort. It is partly for this reason that elephants regularly bathe or cover themselves with mud or layers of dust.

They show a preference for travelling and feeding by night, and during the heat of the day will often spend time in a shady spot, or smear themselves with cool mud if they can find it.

Because of such preferences, sensible and humane owners of captive elephants will not compel their animals to work for too long in hot sunshine, or for days on end in such conditions – as they realise that this kind of forced behaviour is likely to cause deterioration in the animal's health, or resentment in an animal's attitude.

Like their African relatives they are vegetarian in terms of diet and they consume a wide variety of plant matter. Conversely, this seemingly innocuous way of eating often brings about death in elderly animals. Just as plants give life, so too they can be the ultimate cause

Facing page. A wild Asian elephant treading a familiar path. Photo by Rattapol Sirijirasuk.
Following two pages. A wild Asian mother with her baby. Photo by Shutterstock.

ELEPHANT 97

of death. As an elephant reaches old age – and in the wild this may well be after some 50 years or so – a lifetime's consumption of harsh vegetable matter may result in the teeth being badly ground down.

Unable to find enough soft food to sustain himself or herself, the aged animal will simply die of starvation.

Facing page. A captive elephant at Amber, India. Photo by Irene Palmer.
Above. Wild elephants. Photo by Pat Morris.

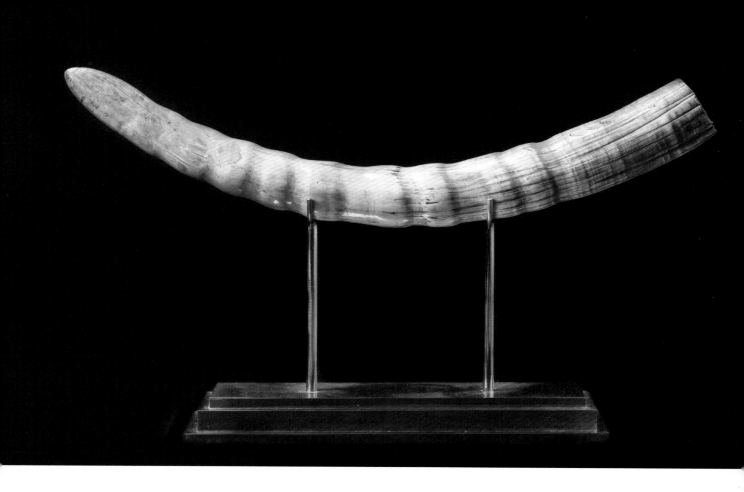

There are, of course, many other ways in which an elephant can die, although natural enemies – apart from humans – are comparatively rare. Tigers are a great danger to the young, but once an elephant is grown, this danger is greatly reduced. Snakebite is another problem, and disease or accident are ever-present risks.

Just as Africa has a mysterious population of 'pygmy' elephants, so too does Asia. These animals, rather smaller than their

Pages 102 and 103. Elephants playing with dung. Photo by Pat Morris.
Above. The slightly deformed tusk of an Asian elephant at the World of Elephants Museum, Sukhothai, Thailand. It is not known whether or not this deformity caused the living animal any discomfort or harm.
Facing page. A Bornean forest elephant feeding and visible for just a few seconds before vanishing rapidly into cover. Photos by John Metcalf.

relatives on the mainland, occur on the great island of Borneo. They are generally regarded as a legitimate subspecies that has been named *Elephas maximus borneensis*, although it is by no means certain that the population is truly indigenous; it may simply result from introduced elephants. The animals occur in the northern and northeastern parts of the island and, as with other elephants, numbers are being reduced due to habitat loss and fragmentation and, of course, the rapidly growing human population, along with the resultant conflict that this creates. Just like African forest elephants, these Bornean creatures can disappear into forest cover with an almost supernatural speed.

Sir John Bowring, a nineteenth-century British diplomat, writing in his book *The Kingdom and People of Siam* (1857), summed up the essence of the elephant in Asia:

> *Without the aid of the elephant it would scarcely be possible to traverse the woods and jungles of Siam. He makes his way as he goes, crushing with his trunk all that resists his progress; over deep morasses or sloughs he drags himself on his knees and belly. When he has to cross a stream, he ascertains the depth by his proboscis, advances slowly, and when he is out of his depth he swims, breathing through his trunk, which is visible when the whole of his body is submerged. He descends into ravines impassable by man, and by the aid of his trunk ascends steep mountains. His ordinary pace is about four to five miles an hour, and he will journey day and night if properly fed. When weary, he strikes the ground with his trunk, making a sound resembling a horn, which announces to his driver that he desires repose.*

Facing page. A small elephant grabs a snack. Photo by John Hodges.

In the Beginning – Evolution, Mammoths, Mastodons

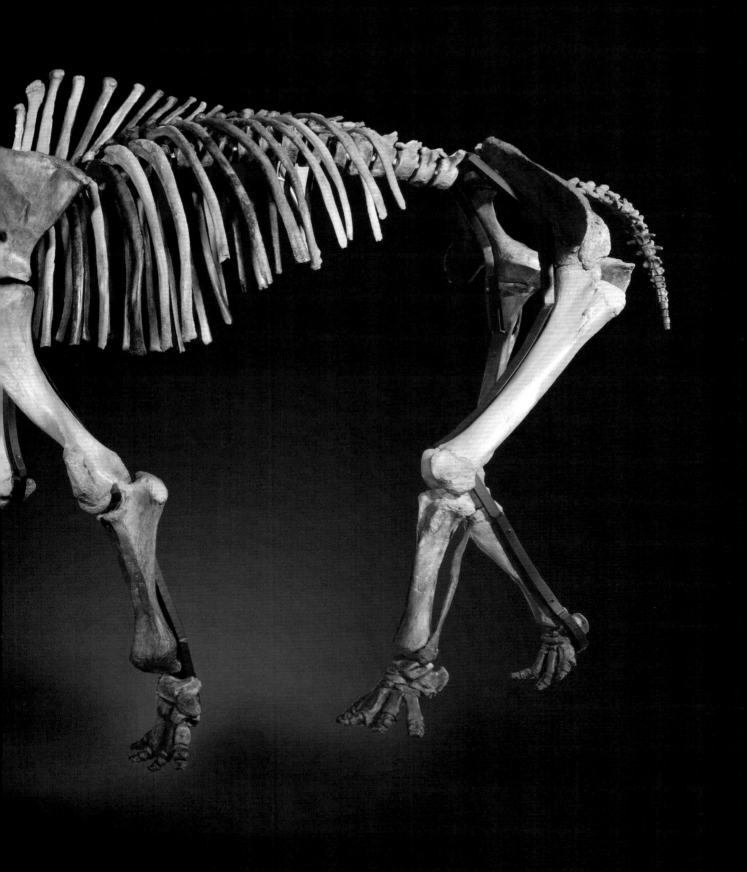

In the Beginning – Evolution, Mammoths, Mastodons

Anyone without specific zoological knowledge who chooses to consider the place of the elephant in the animal world will probably come to the decision that its nearest relatives are the hippos and the rhinos. Such a view would be based on a certain amount of common sense. Rhinos and hippos are the only terrestrial creatures that begin to approach elephants in sheer bulk; they also share a seemingly similar thick, heavily folded skin that lacks any dense hairy covering. But other than the fact that hippos, rhinos and elephants are all placental mammals, they have no real connection with one another.

Indeed, the creatures that are generally considered to be the elephant's closest living relatives seem quite implausible as realistic candidates for this role. They are small, hairy animals that come to attention comparatively rarely and are shaped more like guinea pigs than elephants. The creatures in question are known as hyraxes and there are in fact four different but quite similar species. Various subtle anatomical features connect them with elephants although their ancestral stocks separated many millions of years ago.

Pages 108 and 109. The Summers Place Mammoth. This exceptional mammoth skeleton was found in permafrost alongside a river in Russia, and sold at auction during 2014. Photo by Glyn Clarkson.
Facing page. Dr. Prasert Prasarttong-Osoth standing with a gigantic mammoth skeleton specially prepared for the World of Elephants Museum he founded at Sukhothai, Thailand. Photo by Rattapol Sirijirasuk.

The other animal group with an elephant connection is equally surprising. This is the aquatic family known as sirenians, consisting of the manatees, dugongs and the extinct Steller's sea cow. Some authorities believe that their relationship to elephants may be even closer than that of the hyraxes.

However this may be, the known evolution of the elephant family goes back a long way and although now only the elephants of Asia and Africa still survive, throughout time there have been

many, many different species. Henry Fairfield Osborn (1857–1935), once a fantastically rich and influential palaeontologist, identified and enumerated no less than 352 different species and subspecies that had existed in the past. His rather overzealous attitude has long since been subjected to reappraisal, and today's expert's have rationalized

Above, left. A yellow-spotted hyrax. *Above, right.* A West Indian manatee.
Facing page. The head of a present-day Indian elephant. Millions of years of evolution have led to this remarkable appearance. Photo by John Hodges.

Osborn's work and tend to recognize only half of his designations as valid. But this still represents a remarkably large number of elephantine kinds that have thrived between the Eocene epoch (which began some 65 million years ago, just after the fall of the dinosaurs) and today.

Zoologists place all these similar but different beasts in an order of the animal kingdom known as the Proboscidea – a term that means simply 'animals with trunks'.

The earliest creatures identified as primitive and ancestral elephants are named moeritheres. Their fossil remains are found in certain African deposits of Eocene age, and these fossils show them to have been smallish pig-shaped animals. It is thought that these moeritheres formed the ancestral stock from which all elephant-like creatures are descended.

Just like today's elephants, some of these other descendants have been truly spectacular.

Animals of the genus *Amebelodon* (there are several different species), for instance, were amazing-looking creatures with great shovel-shaped arrangements of the front part of the lower jaw. They flourished during the Miocene epoch and some of their fossils indicate that they were still living around 5 million years ago. *Platybelodon* was a similar-looking creature and its fossils are found quite regularly in parts of Asia.

Facing page. Deinotherium, a spectacular prehistoric elephant that flourished towards the end of the Miocene epoch (about 7 million years ago). This reconstruction shows how the celebrated Czech artist Zdenek Burian imagined and then painted it during 1940.

Pages 116 and 117. A spectacular fossil of *Platybelodon,* a creature that lived in China at around the same time as *Deinotherium* roamed Europe and Asia. This particular specimen is at The World of Elephants Museum at Sukhothai, Thailand, and shows very clearly the extraordinary shovel-shaped lower jaw. Photo by Rattapol Sirijirasuk.

A truly ancient elephant that lived during the Eocene epoch (around 50 million years ago) has been named *Barytherium* and it had short, rather stubby tusks – but there were no less than eight of them!

Coming closer to our own time (and contemporaries of *Amebelodon*), certain gigantic members of the elephant family that flourished have been named *Deinotherium*. There are three known species – one found in fossil deposits of Europe, another in Asia, and a third in Africa. Representatives of these huge creatures may have survived until as recently as a million years ago. Apart from their enormous size, their most remarkable feature was a very strange development of the tusks. These were attached to the lower jaw but pointed downwards and curved inwards. It is not known how these peculiar instruments were used although, naturally, there has been considerable speculation among palaeontologists over their purpose. Some believe they were digging tools employed for rooting in the earth for vegetable matter; others think their main purpose was to break branches so that the animals could reach succulent leaves and shoots.

The dwarf elephants of several Mediterranean islands, most notably Malta and Cyprus, arouse a certain amount of curiosity. Their size ranged from 1.5 metres to 2.5 metres in length and their height was proportionate. How elephants got to these places is a matter of some conjecture, but their fossils show that they did. The most likely explanation is that they were able to gain access during periods when the great ice sheets advanced from the north and sea levels dropped. As sea levels rose again the animals were trapped on relatively small land masses and, once trapped on islands with only limited resources, the forces of evolution resulted in very gradual, but marked, reductions in size.

This is a common phenomenon with mammal populations marooned on islands – survival means that over eons sizes will gradually decrease in accordance with the available food stocks.

Curiously, it is sometimes the opposite way with birds. They will often lose their power of flight and then grow larger and larger! The dodo of Mauritius, descended from an ancestral stock of moderately sized pigeons, is a very good example of this.

The dwarf elephants of the Mediterranean are not the only small elephant kinds that were to be found on islands. Remains of others have been found on various Indonesian islands and also on islands off the coast of California.

The most famous of all prehistoric elephants are, of course, the mammoths and the mastodons. The two words are often thought to be interchangeable, but they are not! From an evolutionary point of view they refer to two different lines of elephantine development.

Yet there are similarities, and these lead inevitably to confusion.

Overleaf. A dramatic view of some of the mammoth fossils on display at The World of Elephants Museum, Sukhothai. Photo by Rattapol Sirijirasuk.

It is perhaps too simplistic to say that the same initial letter 'm' is in any way responsible, but the similarity between their actual scientific names (*Mammut* for the mastodon and *Mammuthus* for the mammoth) is a sure recipe for mixing the two. Add to this, the fact that both became extinct in comparatively recent geological times, and that both were enormous with huge tusks – and the confusion is understandable.

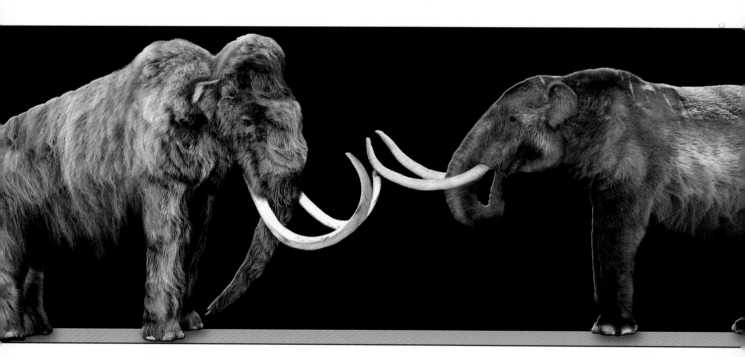

Above. A comparison between mammoth (left) and mastodon (right). The hump of the mammoth and its domed forehead provide a quick means of identification.
Facing page. Flight from the Mammoth (1885). Oil on canvas by Paul-Joseph Jamin (1853–1903), 123 cm x 93 cm. Museum National d'Histoire Naturelle Paris.

As far as is known, mastodons were exclusively North and Central American in their geographical range. There were several distinct species but the best known is *Mammut americanum*, a creature that was widely distributed across the North American continent. It seems that this particular mastodon became extinct surprisingly recently; indeed individuals may have survived until around 10,000 years ago. The reasons for the extinction are probably various but there seems no doubt that humans played a great part in it.

Mammoths too once inhabited North America, having made their way there from Europe across the land bridge that once existed where the Bering Strait now separates Asia from the Americas. Despite the lack of any hard evidence, it is likely that ancestral mastodons also crossed the same land bridge, although this probably occurred during a rather earlier epoch. Whether, and how, the two groups interacted when they met is something that can only be speculated upon.

The word 'mammoth' has, of course, now transcended its original meaning and has become a virtual synonym for anything physically huge or of enormous importance.

Zoologically speaking, it is a rather loose term, for a number of related species can all be described as 'mammoths'. The most ancient is known scientifically as *Mammuthus subplanifrons* and its fossils have been found in Africa. A later ancestral species is

Pages 124 and 125. An evocative image showing a herd of mammoths. Painted and imagined in 1941 by Zdenek Burian (1905–1981), often considered to be the greatest of all painters who specialized in prehistoric scenes.

Above. The mastodon as reconstructed by Charles R. Knight (1874 – 1953), well known for iconic paintings of prehistoric animals. His achievement is all the more remarkable as Knight suffered from seriously impaired vision.

Mammuthus meridionalis, and this form eventually gave rise to the steppe mammoth (*Mammuthus trogontherii*). These were the animals that at some point crossed from Europe and Asia into North America via the land bridge. From this invading stock the Columbian mammoth (*Mammuthus columbi*) evolved and this species may have given rise to others (*Mammuthus imperator* being one of them).

But when people think of 'mammoths' it is the species commonly known as the woolly mammoth (*Mammuthus primigenius*) that is usually under consideration, and this is certainly the mammoth of popular imagination.

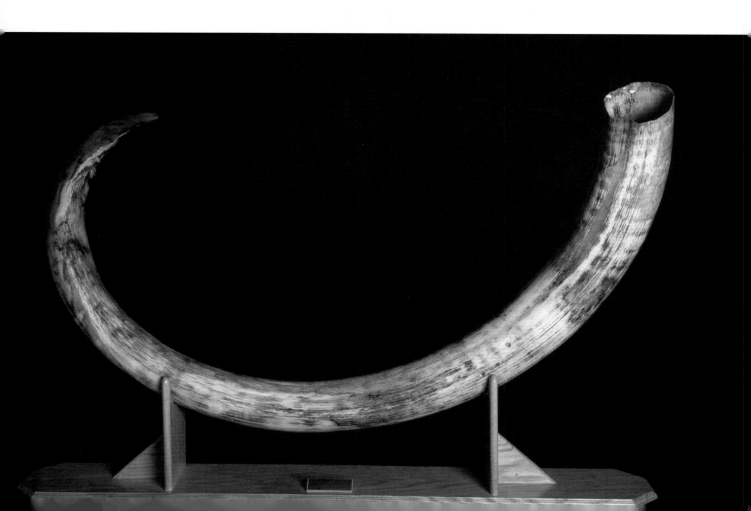

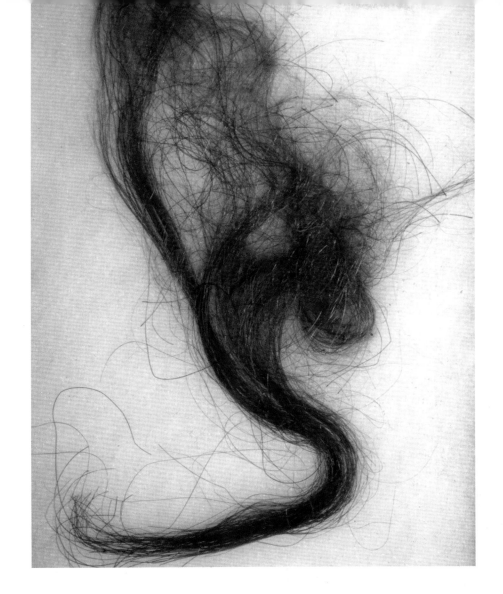

Facing page and above. Mammoth items (tusk and hair) owned by Sheikh Saud Al-Thani, who interested himself in discovering the relics of extinct creatures and formed a formidable collection of such things.

Following two pages. The importance of the mammoth to prehistoric humans. The first image shows a burial using bones and tusks as part of a ritual honouring the dead. It is one of series of pictures of early humans painted by Zdenek Burian (1905–1981). The second is a life-size reconstruction, on display at the World of Elephants Museum at Sukhothai, of a prehistoric shelter using mammoth bones as part of the building process. Photo by Rattapol Sirijirasuk.

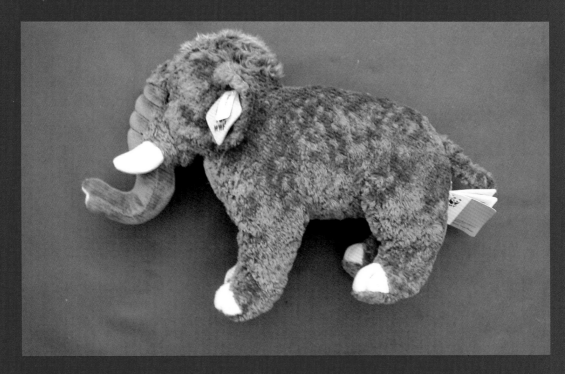

In their heyday – probably around 50,000 years ago – woolly mammoths occupied a vast range of territory from the west coast of Ireland right across Europe and Asia, and on into North America. The species seems to have lived in environments just to the south of the great ice sheets, and there is a consensus of opinion (based on tooth development and the kinds of places in which fossils are found) that these animals were essentially grass eaters – although they certainly supplemented their diet with other forms of vegetation. Why they became extinct is not entirely clear although climatic changes probably contributed, and we can be sure that humans were significant agents in their destruction. One fascinating aspect concerning mammoths is that entire carcases are found in the frozen wastes of Russia, and these discoveries are not especially rare; they are made with surprising regularity.

A remarkable find was the discovery of a specimen now known as the Berezovka Mammoth (excavated during 1901), and another was the finding, in 1977, of a frozen baby given the name Dima. Each discovery caused something of a sensation at the time of its finding. In some ways both are reminders of just how recent the extinction was, and it seems that (in geological terms) the population collapse was fairly rapid. From a seeming population high point of 50,000 years ago, the mammoth was virtually gone by 8,000 years BC.

Previous two pages. Old and New – prehistoric and modern.
First page (above). Drawing of a mammoth on a wall at Rouffignac Cave, France; *(below).* A higly stylized mammoth at La Baume Latrone, France.
Second page. Two twenty-first-century mammoth toys.
Facing page (above). Two views of Dima – the first showing the baby as it was found, the second showing it on a trip to London; *(below).* The celebrated Berezovka Mammoth on display at the Museum of Zoology, St. Petersburg.

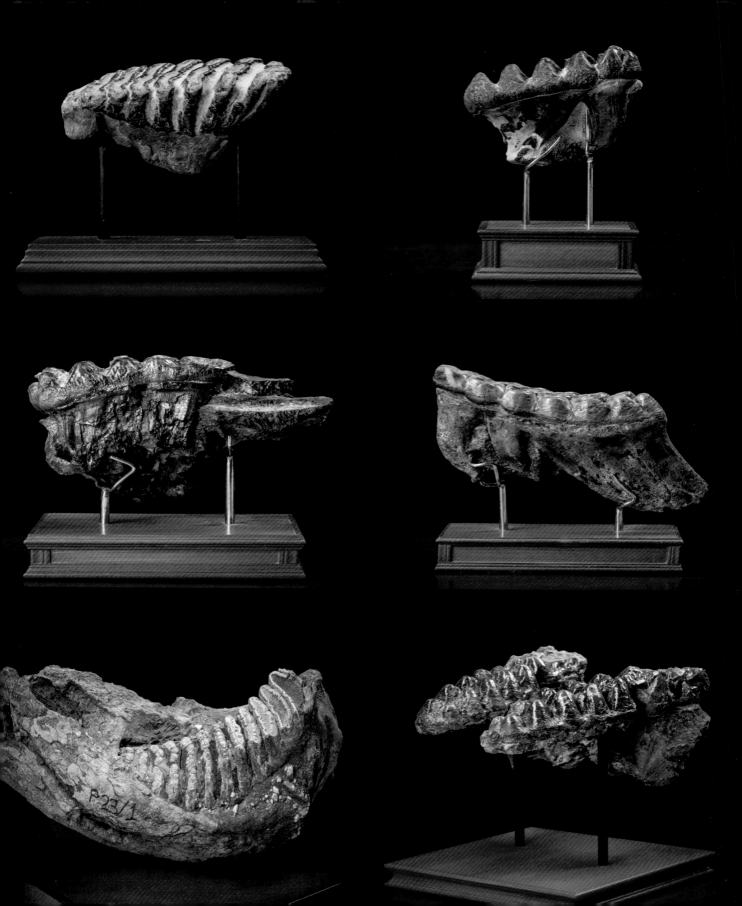

Perhaps a few groups lingered on here and there, but all were gone soon after this time. There is one exception to this rule, however. On Wrangel Island, a small isolated spot in the Chukchi Sea just north of the Bering Strait, a population of mammoths managed to survive for much longer. It is estimated that they held out until around 2,000 BC, so mammoths may have still been roaming the earth (albeit a very remote part of it) when the pyramids of Giza were being built.

Facing page. Teeth from six different species of fossil elephant. Photos by Rattapol Sirijirasuk.
Above. Black tusks from the Miocene of Thailand. Photo by Rattapol Sirijirasuk.
All are on display at the World of Elephants Museum, Sukhothai.

Elephant *Curiosities*

Elephant Curiosities

The legend of an elephant graveyard is one that has long fired the imagination of humans. It hangs on a notion that there is a secret place that African elephants go to when they feel death closing in on them, and that this place contains the bones of hundreds of dead animals – and, if only it could be found, their ivory! Perhaps it is not too cynical to suggest that the reason for the persistence of such a fanciful notion is the wishful thought of tons of ivory just waiting to be found and exploited.

But, unfortunately for those who lust after ivory, there appears to be no real truth in the story, and the whole concept seems nothing more than a myth.

Its origins may lie in the occasional finding of places where two or three individuals have died in close proximity to one another. This may be because old or sick elephants may congregate near water or swampy places where the vegetation is softer, and so die in fairly close proximity to one another. Where this has happened their sun-dried bones, stark and bleached, may litter a considerable area, and due to the creatures' colossal size, there will be an impressive amount of bone material.

Pages 138 and 139. An Asian elephant in the act of painting a self-portrait. Photo taken in Thailand by Rattapol Sirijirasuk.
Facing page. Bones. Photo by Peter Beard.

The legend may also have its roots in the fact that elephants will often show enormous respect for the remains of dead members of their species and will even touch gently and turn over such remains. Leading on from observations of this kind of behaviour, there need be no doubt that elephants do mourn for fallen comrades in ways that are comparable to human grieving.

This is certainly one of the factors that link these animals to human activity and encourage our feelings of empathy towards them. There are many others.

One of these is the controversial matter of elephant painting. First, it must be stressed that Asiatic elephants do actually produce pictures – using their trunks as substitutes for hands with which to hold brushes – in which the subject matter (be it a tree or a flower, etc.) is immediately recognizable. In terms of reality the finished image is rather like one that might be produced by a six-year-old child – crude but easy to identify. The process has been developed as something of a tourist attraction. The elephant paints in front of a group of spectators and the finished picture is sold to one of the group. During this exhibition, or demonstration, the elephant's trainer stands beside his animal, provides it with the tools and paints it needs, and then proceeds to encourage it to perform the required task by means of touches or other subtle signals.

But however it is done (whether through trickery or not) it represents a remarkable achievement that does not seem repeatable with any other animal. Yet various commentators have

Previous two pages. One of the things that endears elephants to humans is the obvious empathy they show, and our ability to recognize much of our own behaviour in theirs. Photo by Pat Morris.

Facing page. Paintings produced by elephants. Photo taken in Thailand by Rattapol Sirijirasuk.

poured a certain amount of scorn on it, suggesting that it is not a display of true creativity, and that if a chimpanzee throws paint at a canvas to produce an abstract image then this shows infinitely more artistic ability. This rather superficial opinion is, of course, largely a reflection of the twentieth-century belief in the sophistication of abstract art, and the assumption that there must automatically be some profound intellectual content behind it.

In true terms of creativity, however, the painting feat that an elephant performs is surely comparable to that achieved by a young child. Just as an elephant may have been trained by tricks learned from his human accomplice, so too a child is encouraged and taught by his or her teacher or parent to produce an image that is recognizable to adults.

Another rather human talent possessed by some elephants is their ability to recognize themselves as individuals when confronted by a mirror. Very few other animals can do this. Apes can, and so too, apparently, can dolphins! Also some members of the crow family seem to have the ability. However, for most creatures this is not something that happens; even those closely associated with humans do not normally appreciate exactly what they are seeing. If, for instance, a dog sees itself in a mirror the general response is either to take no notice at all or else to bark because the animal wrongly senses the presence of an intruder. An elephant, on the other hand, will recognize that the reflection in a mirror, or in water, is indeed an image of himself or herself. Various tests have been devised to check that this is so, the most common being to place a sticker or mark on an animal and then present that animal with a mirror. If the subject of the experiment then looks for the sticker on its actual body it can be assumed that it realises exactly what the mirror image is.

Facing page. Elephants have a keen sense of their individuality and understand exactly how they appear as an individual. Photo by Pat Morris.

Similarly, elephants are capable of gauging the spatial requirements of extraneous objects that have been attached to them. Most animals can, of course, judge with great precision the attributes of their own bodies – whether or not they can fit through a given space, the timing, speed or coordination needed to perform an action, or even its possibility of success. But in most cases this applies only to their own personal body mass or shape. Elephants, on the other hand, can consider and take into account items that have no connection with their own form. For instance, an elephant may innately gauge the size and shape of a howdah that has been loaded onto its back. Often an individual will stop if it feels that the attached howdah will not pass under an overhanging branch standing in its path, sometimes making a judgement call that is dependent on the finest of perceived margins. Then, an elephant may show an additional aspect to its intelligence by its response to its mahout's instructions. If the man taps the offending branch to indicate that all is well, the animal will understand and move on.

There is no doubt that elephants can sometimes form relationships with men, women or children that are truly remarkable.

The tale of Sheila, a young female elephant resident at Belfast Zoo in Northern Ireland during the years of World War II, provides an interesting example.

For various security reasons (mostly connected with the very real fear that dangerous creatures might escape during bombing raids)

Facing page. The relationship between an elephant and its rider can be very close and is sometimes governed by mutual love and respect. Photo by Pat Morris.

many of the zoo animals were slaughtered, but a lady named Denise Austin decided she would take a young elephant home with her – presumably to show that it wasn't a potential danger. Each evening Denise would walk Sheila to her own house, often stopping on the way at a local shop to buy stale bread, a diet supplemented with hay from a local farm; this hay was of much better quality than the sparse food available at the zoo during this period of intense wartime rationing. Sheila would than spend the evening in Denise's garden before being put to bed in the family garage. In the morning she would be walked back to the zoo.

Above. The Belfast elephant with Denise Austin and Denise's mother.

There are other curious stories of elephants in wartime. During World War I an Indian elephant from a travelling menagerie was pressed into service at Sheffield in Yorkshire because all the local horses had been sent away to help with the war effort. She was called Lizzie and was fitted with a harness so that she could help with the moving of heavy metal at a scrap yard. According to one story, she was once caught putting her trunk through an open window to steal someone's dinner.

During the same war, and due once again to the shortage of horses, elephants from a nearby circus were used at Horley, just

Above. Lizzie, the Sheffield elephant.

south of London, to help with ploughing and other heavy work on local farms. The absence of horses was, of course, of great significance at a time when agricultural mechanization was in its infancy.

The much-reported elephant memory is one of the traits that humans can identify with, and there are some remarkable recorded instances that indicate just how powerful it can be. One often retold example was featured in an article in *Scientific American* for January

Above. The Horley elephant helping to work the land during World War I.
Facing page. Do elephants have good memories? Could they perform actions like this if they weren't capable of learning and remembering? Photo from an early twentieth-century encyclopedia showing teak production near Rangoon.

12, 2009, where author James Ritchie recounted the amazing meeting of two female elephants at a sanctuary in Hohenwald, Tennessee. When introduced to one another both animals showed unmistakable signs of extreme excitement and pleasure. In fact they exhibited all the signs of a close former friendship: touching trunks, investigating each other's scars, and so forth.

This remarkable and apparently very intense behaviour caused those who witnessed it to do some research into the previous histories of the pair, and this revealed that some 23 years earlier their paths had indeed crossed. For a period of just a few months they had lived together at a circus.

There is no doubt that from an evolutionary point of view a well-developed memory can be advantageous to creatures such as elephants with highly developed social lives. Obviously, it is particularly helpful for the dominant matriarch in a family group to be able to remember a whole range of things, particularly as her group will follow her lead so readily. Where, for instance, might it be advantageous to go in times of drought or perceived danger? Which specific individuals are likely to be friendly if her group meets another that it has had dealings with in the past?

Research has suggested that an elephant can recognize and remember more than 100 other individuals. Although such abilities are clearly beneficial, in a fast-changing modern world they do not always achieve favourable results. A matriarch may remember a good feeding place visited years previously and lead her group there only to find the situation entirely altered. The area may now be turned over to agriculture or some other human activity – and this, of course, can bring elephants into conflict with local people. And in such a situation there is only one ultimate winner!

One attribute dependent on memory is the desire for revenge, and there are many well-documented instances in which

Facing page. A grainy photo from the 1930s of a performing circus elephant and her two attendants. Do animals such as this one enjoy such antics? It is difficult to say, although there can be no doubt that the training processes used by circuses are cruel and inhumane.

Following two pages. The elephant's ability to form highly developed social groups that rely on qualities such as loyalty, memory, caring for others in the community, and an all-around intelligence is obviously of great importance to survival, but in today's world (increasingly dominated by human activity) it can make them vulnerable and expose them to unexpected danger. The love that these female Asian elephants show so evidently towards the youngest member of their community can leave them open to unscrupulous human behaviour. Photo by Pat Morris.

a domesticated – and seemingly submissive – elephant would bide its time (sometimes for a very long period) before exacting a terrible retribution on a cruel keeper or owner. Usually these revenge tales take a fairly simple form, and the sinned-against animal will pick a suitable moment in which to crush or trample its oppressor or catch him with its trunk and hurl him against a hard object. But one particularly intricate and remarkable tale comes to us from the seventeenth century.

A French travel writer named Simon de la Loubere was visiting Thailand and recounted an incident in which an elephant was punished for some minor misdemeanour. The punishment was bizarre but doesn't seem particularly severe, although it is obvious from subsequent events that the elephant in question thought it was.

The animal was subjected to the indignity (and presumably considerable pain) of having a coconut cracked open on its head. After the cracking was over the elephant calmly collected the broken shell fragments and set them between its front legs, where it kept them for several days, guarding them carefully all the while.

Eventually a moment that the animal deemed suitable arose, and it trampled its tormentor to death. But this was not quite the end of the story, for the elephant gathered carefully the shell fragments it had been guarding and proceeded to place them one by one on the body of the man who had so offended. There is an almost artistic quality about the measured and symbolically balanced elements that feature in this tale of slow retribution.

Facing page. An African elephant at bay. Photo by Granville Davies.

160 ELEPHANT CURIOSITIES

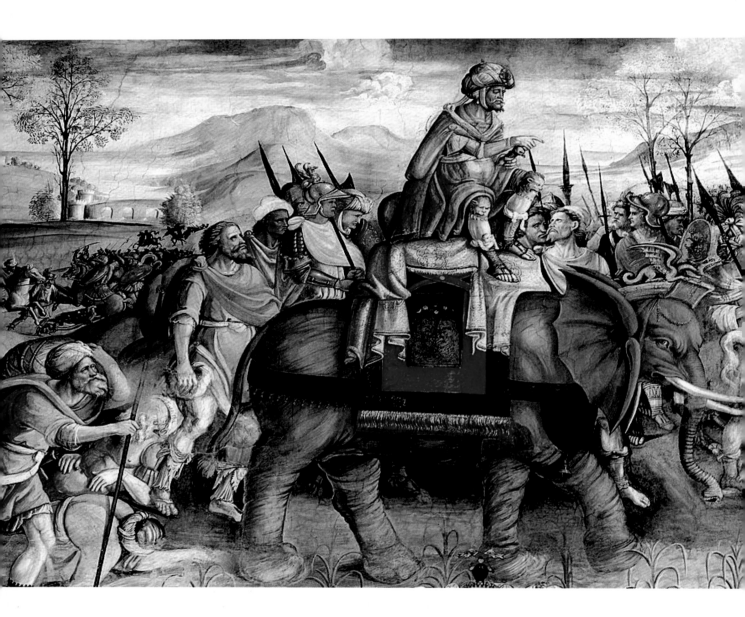

The matter of taming and training is another curiosity that links these animals to humans, and it is clear that Asiatic elephants can be taught to perform all manner of useful, warlike, or even entertaining functions. However, there is a general perception that it is only Asian elephants that are tractable – but this is far from the truth! Although there are comparatively few examples of elephant training in Africa, there are enough of them to suggest that the general lack of domesticated animals is due more to the cultural habits of the people rather than to an unreceptive nature of the animals themselves.

The fact that Hannibal used war elephants from North Africa in his celebrated crossing of the Alps around 218 BC is well known, and obviously these creatures must have been trained to some degree. However, it appears that elephants from Africa had been used by humans even before that. Earlier in the third century BC, Ptolemaic soldiers captured and trained animals near the shores of the Red Sea, and these creatures were then transported by boat into the heartlands of Egypt itself.

Following these endeavours, and then Hannibal's slightly later adventures, there seems to be a gap of more than 2,000 years before any other definite instances are recorded.

Facing page and this page. A fresco in the Palazzo dei Conservatori, Rome, showing Hannibal's advance through Italy after he had taken his war elephants across the Alps. This rather fanciful image of an event that had occurred nearly 2,000 years earlier was produced around the year 1510 by Jacopo Ripanda, a painter who is little remembered today, but one who achieved a considerable reputation during his lifetime. Photo by Jose Luis Ribeiro.

Then, around the year 1880 a group of missionaries based in Gabon managed to capture and train a forest elephant. So remarkable did the indigenous people find this achievement that many travelled for miles to see the results for themselves.

During 1899 the missionaries seem to have come to the attention of a Belgian army officer named Laplume, and he apparently reported what he had seen to his king, Leopold II, a man who regarded The Congo as his personal property and 'garden.' Leopold, well known for his ruthless ability to get things done irrespective of its cost in terms of human suffering, immediately sponsored a programme of elephant training near a village called Api. By the end of the first decade of the twentieth century there were 35 elephants working there – carrying loads, pulling carts, and helping in various ways with cultivation. A second training station was established at Gangala-na-Bodio during the 1920s, but eventually both projects seem to have been abandoned. However, they do indicate that African elephants are by no means the intractable creatures that they have been considered to be.

Since this time, and continuing to today, there are several places in Africa where elephants are used in the service of humans.

Facing page. Three early twentieth-century photographs taken in what was then the Belgian Congo showing African elephants engaging in tasks similar to those regularly performed by elephants in Asia. The taming of elephants in Africa is little known but still occurs today in various parts of the continent.

Above. A trained African elephant being ridden by its keeper. Contrary to popular opinion, African elephants – just like their Asian counterparts – are sometimes trained and used in the service of humans. Photo by David Chancellor.

Above. A bronze Khmer bell now in the World of Elephants Museum at Sukhotai. Found in Cambodia, it was made for a war elephant and dates from the Angkor Wat period (twelfth century). Photo by Rattapol Sirijirasuk.

This page and facing page. The Fable of the Aged Elephant and the Mouse That Hadn't Been Well. A painting by the celebrated artist Raymond Ching. 36 in x 42 in (91 cm x 106 cm), mixed media. Private collection.

An idea that has coloured the opinion of many, and given rise to a multitude of comic situations in cartoons and storybooks, is the notion that elephants are terrified of mice. How this popular idea originated is something of a mystery, but one school of thought inclines to the feeling that it owes something to the vulnerability of the trunk. This incredibly sensitive appendage plays a vital role in almost everything that an elephant does, so individuals are naturally very protective towards it. Obviously, an elephant would be wary of any small, fast-moving creature entering it and causing possible injury. It is noticeable that when sleeping, an elephant will keep the delicate end of its trunk carefully protected and curled up out of harm's way. And certainly, in parts of southeast Asia the manner in which an elephant uses its trunk is thought to be a clear indication of the animal's temperament. In some circles, any individual that habitually swings its trunk from side to side is considered to be a potentially dangerous animal.

To the peoples of India and southeast Asia, the occurrence of an occasional 'white' elephant seems to have an enduring fascination. When one of these aberrant individuals is found it assumes the role of a sacred entity, and in past times ownership usually became the prerogative of kings or rulers. Curiously, the expression 'white elephant' has come to mean something that is essentially useless but cannot easily be disposed of, and paradoxically the phrase derives from the tradition of sacredness. It relates to stories told about historic kings of Siam (now Thailand, of course).

Facing page. A captive white elephant in Thailand. Photo by Shutterstock.
Pages 170 and 171. Two illustrations from a nineteenth-century Parisian magazine *Le Petit Journal,* the first showing the capture of a white elephant, the second showing a war elephant in Thailand.

ELEPHANT 169

Above. The white elephant given to Pope Leo X in 1514. After the elephant's death the Pope commissioned the great Raphael to paint a memorial fresco, a picture that has been destroyed. This drawing, possibly a study by Raphael himself, but more likely a copy of the fresco by Giulio Romano, still survives.

Apparently, if a courtier began to incur the displeasure of his sovereign, he might be presented with one of his king's sacred white elephants. On the surface such a gift represented great honour, and it couldn't be refused. There was another side to the honour, however, for white elephants had to be kept in royal splendour. So ruinously expensive was the upkeep that this was in actuality an entirely unwelcome gift.

One white Asiatic elephant was acquired by King Manuel I of Portugal, and in a display of religious devotion he presented it to Pope Leo X in 1514. It was named Hanno (or Annone) and quickly became a great favourite with Pope Leo. However, it survived under his care for a mere two years, dying while being treated for constipation with a laxative mixed with gold.

So upset was His Holiness that he commissioned Raphael to produce a memorial fresco to which he (the Pope) personally added a few lines of self-penned verse. Unfortunately, the fresco no longer exists, but a copy probably drawn by Giulio Romano still survives.

A white elephant is obviously an abnormality and, as with all creatures, elephants are subject to occasional freak development. Numerous examples of elephants showing physical peculiarities are on record. One unnatural development that has aroused interest at various times is the occasional existence of elephants with four tusks rather than two.

The filmmaker Armand Denis, who was well known during the 1950s and 1960s for his wildlife documentaries, once followed up rumours of such a beast in a particularly remote part of the Ituri Forest.

There, the pygmy inhabitants told stories concerning their belief in a huge animal that lived locally and sported four very long but very slender tusks. In his book *On Safari* (1963) Denis recounts how the creature was eventually found dead by a native hunter and how he (Denis) was able to acquire the skull and tusks – which he supposedly deposited at the American Museum of Natural History in New York.

Museums do, of course, often make use of the spectacular nature of elephant material, and many have displayed enormous stuffed specimens to impressive effect in their entrance foyers. But one museum has taken a rather different approach, and it is a museum entirely devoted to elephants and elephant-related creatures. This museum is at Sukhothai, the ancient capital of Thailand. Called the World of Elephants Museum, it is a reflection of the cultural and historical importance of elephants throughout southeast Asia and the rest of the world, and it houses a vast array of elephant-related material. Yet there is one element in the elephant's story that is missing. There is no emphasis on ivory! The museum's creator and owner, Dr. Prasert Prasarttong-Osoth, is so dedicated to the principles of conservation that he has no wish to promote, or add to, the desirability of such material.

Previous two pages. This spectacular stuffed elephant, produced by the renowned taxidermy company Rowland Ward of London, once stood in the Royal Museum for Central Africa at Tervuren near Brussels but was removed recently. Photo by Annick at Aldo Workshop.

Facing page and two pages following. Three views of the amazing World of Elephants Museum created by Dr. Prasert Prasarttong-Osoth at Sukhothai, Thailand. The museum contains spectacular fossils and life-size models of virtually every species of elephant known to have existed in prehistoric times. Photos by Rattapol Sirijirasuk.

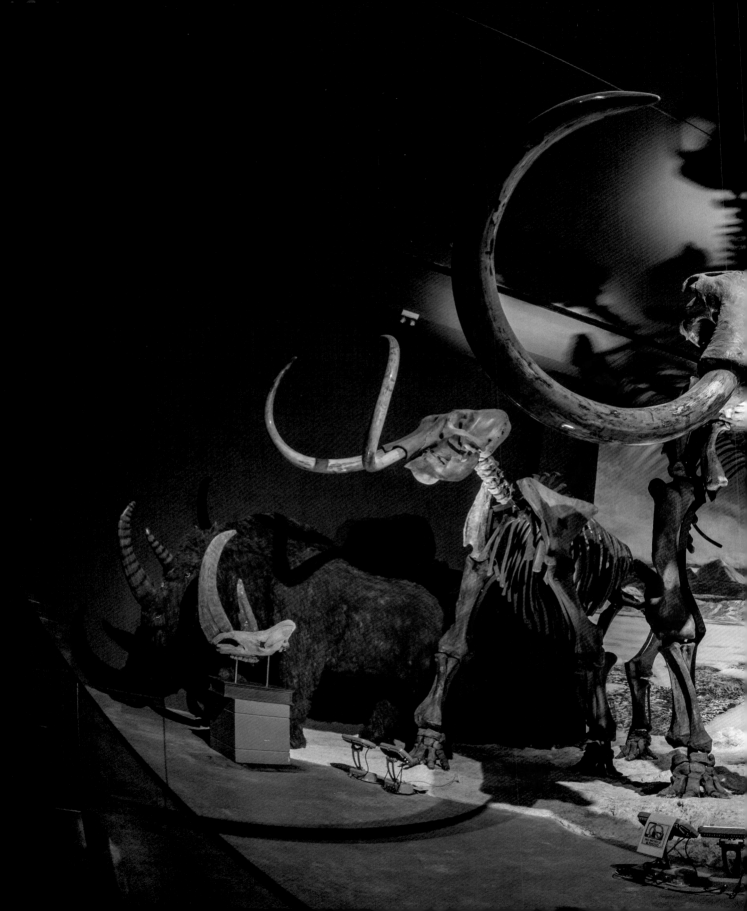

Instead, his museum features life-size models of almost every elephant species known to have existed (prehistoric and modern), an amazing selection of truly spectacular fossil material, and elephant paraphernalia of all kinds. Privately assembled over a lifetime, it is now open to the public and it makes its own curious contribution to the appreciation and love that some humans have for elephants.

Over the years there have been many individual elephants with life stories that stand out for various reasons. Three captives with tales that have resounded down through the years are Chunee, Topsy, and Jumbo. Each tale is distressing, and one is even more repellent than the others, but none show humanity in any of its finer moments.

The earliest of these captive creatures was Chunee, a male elephant imported to London from Bengal in the year 1809. Once arrived he began to make regular appearances on the stage in various theatrical productions, but during 1817 he was bought by Edward Cross, owner of the Exeter Exchange – a once-popular menagerie that occupied a large site on the London thoroughfare still known as The Strand. All seemed well for eight years or so and Chunee continued on a course of considerable celebrity. During November of 1825, however, he fatally gored one of his keepers. Although deemed innocent of the offence (for it was considered an accident) he was fined the sum of a shilling. How this amount was earned or paid is not entirely clear! Just four months later, however, there was another incident. It may have had something to do with his period of musth (something that contemporary accounts appear to allude to as an 'annual paroxysm') or it may have been due to pain from a rotting

Facing page. The execution of Chunee. A contemporary print by George Cruikshank.

DESTRUCTION OF THE FURIOUS ELEPHANT AT EXETER CHANGE

tusk, but his behaviour became difficult and it was impossible to handle him. He began to batter his cage, made from iron and wood, and it was feared he might burst out. Crowds gathered outside just to hear the violent commotions coming from within. Eventually, it was decided that poor Chunee should be put out of his misery. First, there was an attempt to poison him but he refused to take the substance offered even though it was hidden in his hay. Then a group of soldiers were summoned, and they shot him – apparently in a fusillade of more than 150 musket balls. Still the tragic animal did not die, and one of his keepers finally finished him off with a harpoon.

Bad as this story is, the tale of Topsy is particularly repulsive and to twenty-first-century sensibilities it seems almost unbelievable. She was a female Asiatic elephant that arrived in the United States during the last half of the nineteenth century and then spent 25 years or so performing at an American circus.

During these years she acquired something of an ugly reputation and was regarded as having a thoroughly bad nature. This culminated soon after the turn of the century, in 1902, when she killed a spectator.

As a consequence of this action she was sold to Coney Island's Sea Lion Park, where (for what seem to be nothing more or less than commercial reasons) it was decided that she should be executed by hanging.

The full plan was that this event be turned into a public spectacle with tickets being made available (for an admission fee) so that anyone who wished to see the display could view it. The American Society for the Prevention of Cruelty to Animals managed to put a stop to at least some of the voyeuristic elements of these proceedings but was unable to prevent the actual execution and it was carried out on January 4, 1903, before a group of specially invited guests and journalists.

There was a botched attempt to strangle the poor animal using rope attached to a steam-powered winch. Poison and an electrocuting device were also on hand in case the strangulation failed, and eventually it was the electricity that effected Topsy's death. As a final repulsive element to the tale, parts of the execution were filmed by the Edison Company (as a promotional exercise – and this film still exists).

The tale of Jumbo is nowhere near as vile, but even so it has an unhappy ending. It begins around Christmas time in the year 1860 when an elephant was born somewhere in the Sudan. Its mother was killed soon after the birth by a party of hunters and the young animal was captured. After a series of complicated journeys through North Africa and Europe he eventually arrived at the Jardin des Plantes in Paris. The journey was by no means over, however, and he was soon transferred to the London Zoo where he became popular for giving rides to zoo visitors.

Facing page. The electrocution of Topsy in January of 1903.

There the enormous beast (for it had grown to great size) was given the name Jumbo, a word probably derived from 'jambo', the Swahili word for 'hello'. Just like the word 'mammoth', 'Jumbo' has, of course, transcended its original limits. It is no longer just a simple name for a friendly elephant and is now used to describe anything of extra large size.

But despite Jumbo's popularity and the growing use of his name to imply 'huge', a decision was taken to sell him to Phineas T. Barnum (1810–1891), the famous American showman for $10,000 – a vast sum of money by the standards of the time. Appeals were made to stop the sale (even to Queen Victoria) but these were to no avail, and Jumbo crossed the Atlantic.

His arrival caused something of a sensation, much of it due to Barnum's shrewd manipulation of the media. Long before Madison Square Garden staged the fight of the twentieth century between Muhammad Ali and Joe Frazier, Mr. Barnum was showing his celebrity elephant at a previous site for the venue, and the public flocked to see him. One of the peculiar events of Jumbo's life came in 1884 when along with several other elephants he was taken over the Brooklyn Bridge to prove that it was safe enough to take great weight; several years previously, a bridge had collapsed, resulting in a loss of public confidence.

Sadly, Jumbo wasn't destined to live for very much longer. A certain amount of mystery surrounds his death, and several differing versions of the circumstances exist.

Facing page. A promotional poster advertising the arrival of Jumbo in the United States.

What seems likely is that he was being loaded into a container parked on railway lines when an unexpected train came along. Jumbo saw it but couldn't climb the railway embankment to escape and was hit. Barnum, ever ready to capitalize on any event – disastrous or otherwise – gave out a tale that Jumbo had died heroically while trying to shepherd a smaller elephant named Tom Thumb across the tracks. The unscheduled train arrived at just the wrong moment, Barnum said, and hit him. The locomotive was derailed and both elephants died instantly. This, at least, was the showman's version of events. However, another version suggests that Tom Thumb suffered just a broken leg and that Jumbo was already dead by the time the train arrived. The actual truth is, therefore, hidden beneath a series of smokescreens.

Phineas Barnum was not prepared to be left out of pocket by the demise of one of his star exhibits. He had Jumbo's skeleton prepared and his skin stuffed; both were put on display and the skeleton eventually ended up at the American Museum of Natural History in New York, while the stuffed specimen lasted until 1975 when it was wrecked in a fire.

Certain items, including a British policeman's whistle, had been found when Jumbo's stomach was dissected, and these were put on exhibition. Curiously, his celebrity didn't completely dim over the years, and some of the ashes were saved after the fire that destroyed his stuffed remains – and these ashes still exist!

Facing page. Jumbo in death.

Because a certain amount of spirituality seems to surround the ways of elephants, perhaps it is appropriate to finish this chapter on a more mystical note. A conservationist named Lawrence Anthony was well known for the fine work he did with animals of all kinds – but most particularly for the efforts he made with rhinos and elephants. He is probably best remembered for his popular book *The Elephant Whisperer* (2009). One of his greatest achievements was to save, and then care for, an entire herd of very unruly elephants, something he did by first placing himself in an enormous amount of personal

Above and facing page. The once-unruly herd of elephants that arrived at the home of their saviour Lawrence Anthony soon after his death, and had seemingly come to mourn his passing. These animals are said to have made an unbidden 12-hour journey through the bush just to be there, and stand placidly behind the fence that protected his house. After a day or so they left, but a year later they returned and loitered around aimlessly once again. Photos reproduced by permission of Thula Thula Game Reserve.

danger. Some years after his encounter with these animals, during March of 2012, Mr. Anthony died quite suddenly of a heart attack, and the day after his untimely death something very strange happened.

The elephants he had saved lived some distance from his house; in fact it was a 12-hour march away, a march through difficult bush country. Yet, quite unbidden, his herd of elephants arrived

ELEPHANT CURIOSITIES

outside his house having made a very deliberate trek to get there. The next day a second herd arrived and both groups just loitered, seemingly aimlessly. After a day or so, they wandered back into the bush.

But this is not quite the end of the story. One year later, apparently to the day, they all returned and again just stood around for a while before eventually wandering off. Another year passed, and once again on the anniversary of Mr. Lawrence's death, the elephants returned.

No one knows why. Perhaps it was all just a coincidence and nothing more! Who knows?

Facing page. There are many instances and stories of elephants showing emotions or behaviour that men and women can relate to, and there are many photographs that illustrate the same thing. The Thula Thula tale may be one example of this. The photo opposite, taken in Holland at the Dierenpark, Emmen, is certainly another.

The Elephant in Art, Literature, and Popular Culture

The Elephant *in* Art, *Li*terature, *an*d *Po*pular Culture

One moonlit night a little over two and a half thousand years ago a woman sat forlornly by a lake somewhere in the Himalayas. She longed for a baby but had remained childless for 20 years. Soon she fell asleep, and then she dreamed a dream. An enormous white elephant seemed to come from nowhere and it was carrying in its trunk an exquisitely beautiful white lotus flower. The great animal circled around her three times, then magically entered her womb by the right side. When the woman awoke, she found herself alone upon the cold lake's shore and lingered for a while. But the elephant didn't come back, even though she waited. However, something marvellous did happen.

Pages 192 and 193. The 'Ma-Robert' and Elephant in the Shallows, Shire River, Lower Zambezi (1859). Thomas Baines, oil on canvas, 17 in x 26 in (44 cm x 65 cm). Royal Geographical Society.

Facing page. Wat Sorasak, an ancient memorial celebrating elephants among the Buddist temples at Sukhothai, once the capital of Thailand.

Ten months later, to the very day, she gave birth to a son, a son who was immediately able to walk. But, naturally, the soles of his feet were tender, and to give protection lotus blooms opened beneath them. The child was to become known to the world as the Buddha, the lotus was to become a symbol of wisdom and purity, and the elephant was to become a symbol of true greatness.

This story may be the elephant's first important entry into literature (although there are one or two earlier mentions), but perhaps its most curious element is the great phonetic similarity between the name of the lady beside the lake, Mata Maya, and Mother Mary. Does it mean anything? Perhaps. Perhaps not.

As anyone who has travelled in India or southeast Asia knows, the association with Buddha is by no means the only element in the reverential attitude that the inhabitants have for elephants – even though that reverence is stained and contradicted by the appalling cruelty that is regularly shown towards these creatures. The intense regard has occurred throughout the ages and continues in all sorts of forms to the present day.

Facing page and above. Two ancient and monumental elephant sculptures, typical of many in southern Asia. These two are at Khajuraho, Madhya Pradesh, India. Photos by Irene Palmer.

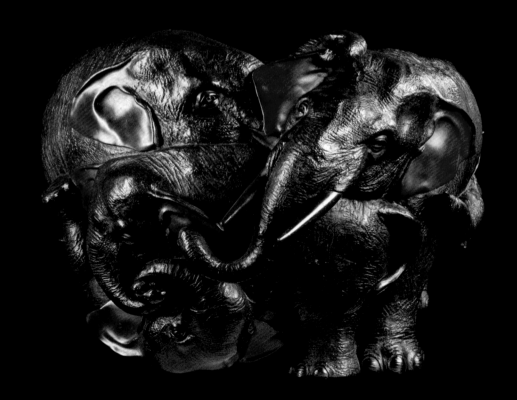

Previous two pages. Four antique pieces featuring elephants. *Clockwise from top left hand corner:* An eighteenth-century gilt copper repousse elephant head. A bronze inkwell with gilded trappings and ivory tusks in the form of an elephant head (French, circa 1880). Both of these images courtesy of Peter Petrou, London. A Chinese green glazed ceramic from the Han Dynasty, 200 BC to 220 AD. Courtesy of the World of Elephants Museum, Sukhothai; photo by Rattapol Sirijirasuk. A bronze Japanese jardiniere from the Meiji Period (1868 – 1912). Courtesy of Peter Petrou.

Throughout the region, depictions of elephants (both in sculptural form and in painted images) proliferate and these are usually held in high regard.

The image, and indeed the whole concept, of the elephant is of vital importance in religion, cultural celebration, and all kinds of sacred ceremonies.

There are many, many examples, of its preeminence – one of the best known being Ganesh (or Ganesha), the Hindu elephant god with four arms. Appropriately, perhaps, he is associated with wisdom, success, and good luck.

In ancient Europe, the best-known early story concerning elephants relates to Hannibal's crossing of the Alps, but another tale that comes down to us from slightly later is the story of Eleazar Maccabeus, a Hebrew hero whose exploits are related in some editions of *The Apocrypha*, and whose tale was more familiar in past centuries; in fact it often served as the basis for elaborate illustrations.

Around 160 BC Eleazar went into battle against an enemy king, Antiochus V, who was apparently mounted on a war elephant. As might be expected, the elephant was carrying all before him, so Eleazar lay on the ground and speared the creature in the side, killing it. However, as it fell it landed on its attacker and killed him in the process!

Facing page. Ganesh, bringer of good luck, the Hindu elephant god with four arms and two legs.
Following two pages. Two illustrations showing the heroic death of Eleazar. The first is an engraving made by by Charles Heath in 1815 for the Macklin Bible (after a painting by Philip James de Loutherbourg (1740 – 1812)). The second is from a mediaeval manuscript known as the *Speculum humanae salvationis* (or *Mirror of Human Salvation*), which dates from the first decades of the fourteenth-century.

HEROISM OF ELEAZAR.

1. Book of Maccabees Ch. 6. v. 43. 44. 45. 46.

The elephant's first appearance in a printed book entirely devoted to the subject of elephants comes much, much later, of course. *Elephantographia Curiosa* was published by a Leipzig physician named D. Georg Christoph Petri von Hartenfels in 1715, with a second edition released seven years later. Twenty-eight richly detailed copperplate engravings by German artist Jacob Petrus feature elephants in their natural habitat, in historical events, in battle or in play and help to make this an intriguing work. The text details elephant anatomy and the differences between the Asian and African species, as well as highlighting the creatures' importance in many areas of human life.

Above, and facing page. Two illustrations from *Elephantographia Curiosa*, and the title page.

D. GEORG. CHRISTOPH. PETRI
ab Hartenfelß,
CONSILIARII ET ARCHIATRI MO-
GUNTINI &c. &c.

ELEPHANTOGRAPHIA CURIOSA,

EDITIO ALTERA AUCTIOR ET EMENDATIOR, CUM MULTIS FIGURIS ÆNEIS,
CUI ACCESSIT

EJUSDEM AUCTORIS ORATIO PANEGYRICA DE ELEPHANTIS, PUBLICE IN ACTU DOCTORALI
ERFORDIÆ HABITA;
NEC NON

JUSTI LIPSII EPISTOLA DE
EODEM ARGUMENTO ERUDITE
CONSCRIPTA:
IN FINE ADJECTUS EST INDEX RERUM NO-
TABILIUM LOCUPLETISSIMUS.

LIPSIÆ & ERFORDIÆ,
Typis & impensis JOH. MICH. FUNCKII,
M DCC XXIII.

Magnos magna decent. *Aut mors aut vita decora.*

ELEPHANTOGRAPHIA
Curiosa
D. Georgii Christophori
Petri ab Hartenfels.

Manet integra Laurus. *Nec hydra nocebit.*

Pura Placent Superis. *Mansuetis grandia cedunt.*

J. J. Hiltebrandt delin.   Jacob Petrus Sculpsit Erffurti.

Since this time, as might be expected, there have been many books on elephants, and during the last 50 years or so there has been a publication explosion with literally hundreds of factual volumes on all aspects of elephant behaviour and legend. Many of these have enhanced our knowledge of elephants and elephant behaviour in the

most profound ways. One of the most important and influential is Peter Beard's *End of the Game*, first published in 1963 but republished several times since then. It is a spectacular and tragic visual record of human involvement with elephants.

Facing page. The frontispiece for *Elephantographia Curiosa* (1715).
Above. The cover image for an edition of Peter Beard's *End of the Game* (1963).

Few of these factual memoirs of elephants take into account all of the thousands of childrens' books and other genres that use elephants as central characters. Fictional elephants played a role in Rudyard Kipling's famous *Jungle Book* (1894), a work that much later inspired the movie of the same name.

Among the first of the well-known children's books that feature imaginary elephants was *Babar the Elephant*, a creation published in France during 1931. Its author is Jean de Brunhoff, but the story began in bedtime tales told by his wife, Cecile, to their children and then developed by Jean (who was already an illustrator). His cartoon creature became an instant hit, quickly becoming an international success with the original concept spawning a long series of *Babar* titles. De Brunhoff died suddenly aged just 37, but one of his sons developed a similar drawing style and continued the series, largely to keep the memory of his father alive.

The list of storybook elephants is virtually endless, one of the most popular being *Mumfie*, written by Katherine Tozer; as with *Babar*, the original idea led to a series.

Perhaps the cartoon elephant found its most famous expression in *Dumbo*, Walt Disney's animated feature film. First distributed in the United States during October 1941 (just before the US entered World War II), it features a cartoon elephant with enormous ears that in the story give rise to ridicule – but in actuality allow the central character to fly.

Facing page. The cover and spine of an early edition of *The Jungle Book*.
Page 210. The cover of an edition of *The Story of Babar*.
Page 211. Four illustrations from *Mumfie*.

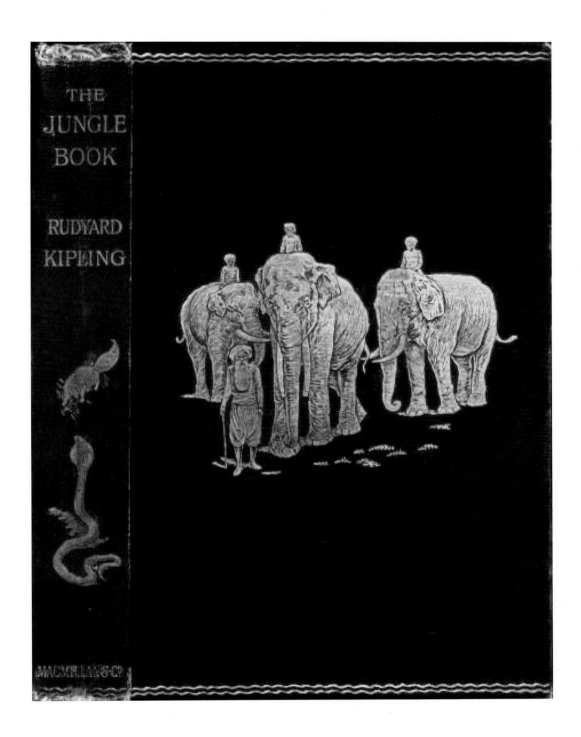

Above. The Triumphs of Caesar by Andrea Mantegna, 105 in x 109 in (262 cm x 278 cm), tempera on canvas. One of a series of nine huge paintings produced between 1484 and 1492 for the Gonzaga Ducal Palace, Mantua; the nine pictures are considered one of the great masterpieces of the Renaissance. They were all acquired by King Charles I in 1629, and unlike most of his other possessions were never sold off by Oliver Cromwell's revolutionary government when it came into power after the king's fall. They now form part of the Royal Collection at Hampton Court Palace, London.

As far as an appearance in Western art is concerned, the tradition begins with cave paintings of mammoths. Perhaps the most spectacular series of such images is at Rouffignac in the Dordogne area of France, where one particular chamber is known as The Cave of 100 Mammoths (*see* page 132). The actual date of the paintings is uncertain but they are thought to be at least 12,000 years old. Other cave systems in Europe contain impressive mammoth paintings, but far less well known are petroglyphs (images made by scraping, incising, carving or abrading) of either mammoths or mastodons (specific identity cannot be determined) hammered onto rock faces of the Colorado Plateau in the southwestern United States.

After the mammoth's extinction, the whole notion of an elephantine creature became almost legendary; certainly it was remote and fascinating to Europeans. Hannibal's crossing of the Alps with elephants and subsequent invasion of Italy passed into myth, and during mediaeval times the occasional appearance of an elephant (brought with considerable difficulty from Africa or Asia) usually caused something of a sensation. The event was often depicted in primitive fashion by a local artist or monk.

The Renaissance heralded a greater sophistication in artistic representation, but as far as elephants were concerned the representations generally harked back to Hannibal or subsequent Roman successes. One of the great masterpieces of Renaissance art is *The Triumphs of Caesar*, painted by Andrea Mantegna between 1484 and 1492. This massive work consists of nine large panels that celebrate the victory of Julius Caesar during the Gallic Wars. The fifth panel in this monumental series shows a triumphal procession featuring elephants and is arguably the first important Renaissance depiction of these animals. Although it began life in Italy, the series was purchased by King Charles I in 1629 and removed to Hampton Court in London, where the paintings have remained ever since.

214 THE ELEPHANT IN ART, LITERATURE, AND POPULAR CULTURE

After the king's fall and subsequent execution most of his possessions were sold by Parliament. Very few objects survived this cull, but the Mantegna series was considered to be so important that if a sale was ever considered, such an action was quickly overruled and the paintings remained where they had been placed by the king.

The influence of the series was so great that in 1630, very soon after their arrival in London, Peter Paul Rubens was inspired to paint

Above. A Roman Triumph by Peter Paul Rubens (circa 1630), 33 in x 64 in (86 cm x 163 cm), oil on canvas stuck down on oak panel. National Gallery, London.
Facing page. Detail from the central panel of *The Garden of Earthly Delights* by Hieronymus Bosch (circa 1500). Museo del Prado, Madrid.

A Roman Triumph, a picture that now hangs in London's National Gallery. From the elephants' point of view these 'triumphs' that the pictures glorify ended badly. The captured animals were, apparently, eventually killed in Roman circuses.

Reflecting the darker side of human activity with elephants, it is inevitable that Hieronymus Bosch should have turned his attention in that direction. Bosch seems to have seen an actual elephant, for he included one in the background of one of his panels for his great work *The Garden of Earthly Delights*. Whether he

produced another painting, this one showing a war elephant, or whether he simply made a drawing is not known, but whatever this image may have been, it has been lost and only a copy (produced in several variants) remains. The elephant carries a Turkish insignia so presumably the picture reflects European fears that an invasion from the East was likely.

Giuseppe Arcimboldo, the painter of so many eccentric images in the last half of the sixteenth century, also could not resist including an elephant in one of his fanciful constructions. In 1566 he painted four panels depicting *The Four Elements,* and one of them, *Earth*, shows a head, formed in his characteristic style, with an elephant as one of its essential ingredients.

Above. An image produced by an anonymous hand after a lost original by Hieronymus Bosch.
Facing page. Earth by Giuseppe Arcimboldo, 1566, 18 in x 27 in (48 cm x 70 cm), oil on panel. Private collection.

ELEPHANT 217

Just a little after Rubens had constructed his painting of *A Roman Triumph*, Rembrandt produced (around 1637) another picture that is now in London – a drawing made in black chalk and charcoal. Not only is this picture important for its artistic virtue, it also represents a significant milestone in the story of zoological science, as it appears to be a representation of the actual individual first used to distinguish the Asiatic elephant as a species.

The model for the drawing was a tame female elephant known as Hansken, and she became something of a celebrity in Holland at around the time when Rembrandt encountered her.

Apparently, she could wave a sword or a flag, holding these and similar items with her trunk.

Another trick was to place a hat on her own head. Such antics endeared her to the public and ensured her popularity.

After her death her skeleton was preserved and eventually found its way to the Natural History Museum, Florence – where it is still on display! It is this skeleton that forms the type specimen (a term used by zoologists to identify the original example used in the first scientific description of a species) for the species *Elephas maximus*.

Facing page and above. Two views of Hansken, the Dutch elephant. The illustration opposite is a contemporary engraving showing Hansken and some of the antics she got up to. The drawing above is by Rembrandt (circa 1637) and was made using black chalk and charcoal on paper. 7 in x 10 in (18 cm x 26 cm). British Museum.

Along with painted or drawn images of elephants there are, of course, pieces of ancient art made from the one part of the elephant that was prized above all others – ivory. Sometimes the image of the animal itself was carved but more often the material was used for other subjects.

Above. A very old carving of Ganesh made from Asian elephant ivory. This ancient relic has lost two of its limbs over time.
Facing page. An incredibly beautiful ivory carving of the crucified Jesus, probably made in Italy during the late sixteenthth century.

Two utilitarian items made from Asian ivory.
Above. A fantastically ornate and elaborate howdah probably made in the nineteenth century and now in the Bangkok National Museum.
Facing page. A delicately made antique ivory pipe now in the World of Elephants Museum, Sukhothai. Photo by Rattapol Sirijirasuk.

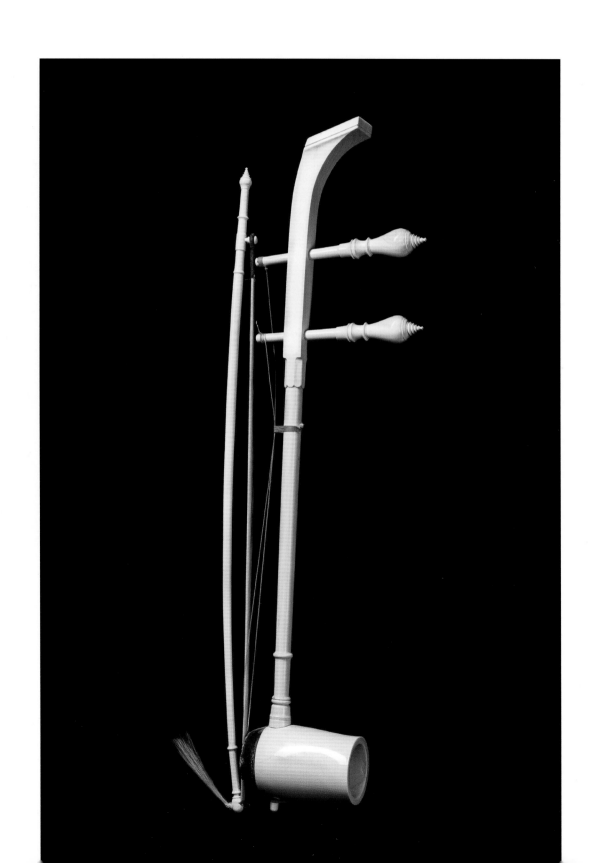

The Elephant of the Bastille. The final design for a fountain that was erected in Paris on the site of the Bastille and stood there between 1813 and 1846. Aquarelle by Jean-Antoine Alavoine (1776–1834), 16 in x 20 in (41 cm x 51 cm). Musee du Louvre, Paris.

It is perhaps surprising to find that the visual image of the elephant did not figure more largely in the art of the Renaissance and the following years. There is, for instance, no image with the same iconic status as Durer's famous picture of a rhinoceros. And this seems to be a trend that continued. After Rembrandt's drawing of the late 1630s there is a dearth of elephants in art. Certainly, none of the major artists produced any pictures of particular interest.

It is not until the nineteenth century that artists began to feature elephants in a significant way.

A fascinating elephant depiction of this later period was made by the relatively little-known artist Thomas Baines, who in 1859 painted *The Ma-Robert and Elephant in the Shallows of the Shire River, Lower Zambezi, Mozambique*. A wounded elephant raises its head, trunk, and tail in defiance as it faces a fusillade of shots from a steam launch with the rather peculiar name of *Ma-Robert*. Baines had been on board this steamer with David Livingstone as it proceeded up the Zambezi River (although they subsequently quarrelled), and the artist wished to record an incident that symbolized the meeting of European technology with animals from the virgin wilds of Africa.

But even in the nineteenth century it was illustrators producing images for books, magazines, and journals who were depicting elephants – rather than those who considered themselves exponents of fine art.

Previous two pages. Elephant art from different places and different times.
First page (above). A typical sixteenth-century book illustration; *(below).* A Dutch or Flemish engraving by or after Gerard Groenning (circa 1565).
Second page (above). An elephant with rider. Amal-e Hashim (circa 1640), 6 in x 8 in (15 cm x 20 cm). Ink and watercolour on paper (badly water damaged). Metropolitan Museum of Art, New York; *(below).* The elephant paddock and wapiti house at the London Zoo. Lithograph by G. Scharf (1835).
Facing page. A typical 1930s poster advertising the London Zoo by K. Nixon.

There were exceptions like Joseph Wolf or Wilhelm Kuhnert, but even these are often considered simply 'animal painters'. And this is a trend that continued into the twentieth century, when it was often illustrators operating in the field of advertising who featured elephants, or specialists in what has come to be known as 'wildlife art'. Painters with more general intent have largely left the subject alone.

Sculpture is perhaps a different matter and the animalier bronzes by exponents like Antoine-Louis Barye and, of course, later sculptures by Rembrant Bugatti are well known.

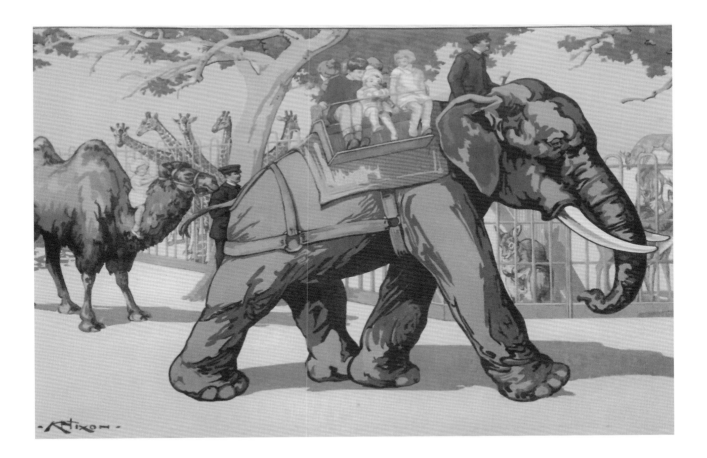

A nineteenth-century sepia image, artist unknown, titled *Return from Hog-hunting*.

There are, of course, exceptions to generalizations, and a few painters have used the elephant motif in imaginative ways. Salvador Dali, for instance, featured elephants several times in his typical surrealistic manner. But most others, if they have used elephants at all, have included them as decoration instead of seeking to make any profound point.

The images that predominate are those made by practitioners who specialize exclusively in the genre of wildlife art, and the most notable exponent in this field is the English painter David Shepherd who has spawned a host of followers and outright imitators.

The problem that exponents of this genre have is an obvious one. They have been entirely overtaken by photography. There are now so many wonderful, evocative photographic images of elephants in all sorts of situations, and many of these make truly profound points. It is virtually impossible for those operating in the wildlife art genre to make any meaningful statement that has not already been better expressed through photography. And perhaps this is also the reason why painters with a somewhat wider remit don't often attempt the task.

Among contemporary artists only a very few have done so, and tried to feature elephants in a fresh way. Peter Beard has achieved this with his collages, and Raymond Ching (see pages 26 and 166) and the American artist Walton Ford have tried to express the idea of the elephant in an expressive manner, both showing nods in the direction of popular culture.

Facing page. Wilhelm Kuhnert (1861 – 1926) was an early exponent of what might be called wildlife art, producing many easel paintings of elephants, and images for books and journals.

Previous two pages. Nila. Walton Ford (1999 – 2000), watercolour, gouache, pencil, and ink on paper, 144 in x 216 in (365 cm x 548 cm). Courtesy of Walton Ford and Paul Kasmin Gallery. © Walton Ford.

This page and facing page. It is virtually impossible for a modern wildlife painter to express the power, grace, and beauty of the elephant in a way that compares favourably with the effects that skilled and patient photographers are able to achieve. The stunning and beautiful potential of photography is displayed in this wonderful image captured by David Chancellor.

This page and facing page. Like Raymond Ching and Walton Ford, the Thai artist Prateep Kochabua has approached painting elephants in a rather different manner to most contemporary painters, and has often used their image to embellish his remarkable visual fantasies. This one is called *Engaged in Combat* and was painted in 2012. Oil on canvas, 94 in x 118 in (240 cm x 300 cm). Reproduced by kind permission of the artist.

As far as advertising or promotional material is concerned, the position is rather different, and a number of artists have produced truly striking images. The strange shape, enormous size, and visual familiarity of elephants all contribute to the production of graphic images, something that illustrators have taken full advantage of. Many of these kinds of images – particularly those made during the first half of the twentieth century – were created to promote circuses or zoos. More recently they tend to promote products or companies, usually emphasizing power, strength, size, or even lovability as part of the advertising process.

Above. Circus elephants.
Facing page. A particularly graphic illustration advertising the London Zoo and the London Underground system.

GIL GRAY CIRCUS ELEPHANT BALLET

Facing page. A typical zoo poster from 1935.
Above. Are the elephants also having fun? Perhaps.

Facing page. Did these elephants really play tunes as the poster suggests? Who knows?

Above. A contemporary use of the image of the elephant in advertising. There is a certain irony in the association of elephants with the suggested reduction of human impact on the environment.

Today, the idea of the elephant in the popular imagination is dominated by two connected ideas – that of the appalling cruelty to which these creatures are subjected, and their general plight. More than ever before there is a general revulsion towards the persecution of elephants for their ivory, along with the hideous conditions in which captive elephants are often kept and the cruelty they are exposed to. Whether this increasing awareness can overcome vested interest, and the increasing burden that human overpopulation places on the wild places of the world, remains to be seen.

Facing page. Detail from *The 'Ma-Robert' and Elephant in the Shallows, Shire River, Lower Zambezi* (shown in full on pages 192 and 193). In 1859 when Thomas Baines (1820 – 1875) painted this picture, he intended to show the disastrous consequences of the meeting of wild African animals with technological humans. His elephant stands at bay, trunk and tail raised in defiance, against the fusillade of shots fired at it by those on board the *Ma-Robert*. Did Baines realize just how prophetic his painting was? Maybe. But perhaps even he did not realize the awful extent of the damage that humans were destined to effect. Baines's evocative painting is now the property of The Royal Geographic Society.

Conservation

Conservation

*When I first escaped to East Africa in August 1955...
it was one of the heaviest wildlife areas...in the world...
No one then could have guessed what was going to happen.
Kenya's modest population of five million...suddenly
became a starving population of over 30,000,000...
Millions of years of evolutionary processes were interfered
with, cut down, fenced off, shot out, sub-divided. The
Pleistocene gets paved over and this is the End of the
Game...We cunningly adapt to the damage we cause in
a big blame game of politically correct spin....teeming
populations unchecked, uneducated leadership, crude oil
and crude politics, tribalism and territorialism, sensory
saturation, conspicuous consumption ...anarchy, insurgency,
war, disease, mad cowboy, historical mediocrity, betrayal
and shame...Let's just welcome it all and take notes while
the world destroys itself.*

Peter Beard (2006)

Pages 248 and 249. Elephants casting shadows. Photo by Tim Flack.
Facing page. An African bull elephant. Photo by Granville Davies.

This chapter is written in a spirit of hope but it is coloured – even overshadowed – by the sentiments just expressed. The words were written in 2006 for the introduction of Peter Beard's astonishing eponymous book, but if anything they are even more true today than they were then.

The reality is that the problem of elephant conservation is a complex one.

In many quarters it has been reduced to the single issue of ivory poaching and its elimination, but in truth this is just part of a much, much bigger issue. The horrors of ivory hunting are self-evident, but the despairing words of Peter Beard underline the full scale of the problem, both for elephants and for humans.

Men, women and children are generally – and immediately – attracted to elephants, and there are all sorts of reasons for this. The more these amazing creatures are studied, the more apparent it becomes that they share many traits and similarities (in behaviour, social structures, emotions, etc.) with us. These kinds of factors, plus their enormous size and presence – and often their friendliness and usefulness – have endeared elephants to humans for centuries.

And yet this fairly universal admiration has created a tragic contradiction.

Facing page. A herd of elephants raising dust as they search for food in a fairly barren environment. Photo by Granville Davies.

While elephants have been loved, respected, valued – even revered in many places – one particular part of their anatomy has resulted in horrendous cruelty and slaughter, a slaughter that, sadly, shows little sign of abating. The huge size, beauty, and spectacular appearance of the tusks has made them an obvious target for humans. Once it was found that elephant ivory had a particularly fine grain, it became apparent that it would serve as a very suitable substance for making intricate carvings.

Facing page. An illustration produced by Gilbert Holiday (1879 – 1937) for *The Graphic* magazine (March, 1909) titled *The Ivory Merchants at the London Docks*. The piles of tusks, the acute interest displayed by the men, and the enormous weighing equipment indicate the vast scale on which ivory has been traded year in, year out. And this is a scene at just one venue in one city, in one year; the reality is that scenes of this kind still are taking place in certain parts of the world.

Above. The grim reality of ivory hunting.

Unlike bone, ivory is without blood vessels and this too makes any carefully wrought carving likely to be more pleasing. Add to this the fact that ivory has a wonderful lustre that increases with age, and its desirability increases still more.

For centuries the tusks have been in high demand even in places thousands of miles from the areas that elephants inhabited, and in periods of time when the actual creatures from which the tusks derived were no more than figments of the imagination and of legend.

The tusks from dead elephants were carried manually for vast distances from the interior of Africa to its coasts and then transported to far distant lands, and all this at periods when movement was difficult and mainly unaided by mechanical forms of transport. The names of places, even of entire countries (for instance the Ivory Coast), bear witness to this trade so hideous in its implications. And there is no real letup in the disgusting killing of elephants for their tusks. Some of the estimates of the number of elephants that are killed for their ivory every year are truly horrifying.

But in a rapidly changing world the persecution of elephants is no longer simply about ivory. Now it is a much more complex problem, and one that spells an even bigger disaster for elephant populations.

Facing page. Asian elephant tusks carved during the late nineteenth century. Even a deformed and twisted tusk was used to produce this intricate work. Although decorative work of this kind is no longer fashionable or wanted in the West, it is still highly sought after in parts of the Far East, and items very similar to this are still being produced.

Pages 258 and 259. Two African elephants, alone and vulnerable. Photograph by Granville Davies.

The simple fact is that for those who live close to wild elephants, the question of compatibility arises. Ever-increasing human populations bring the interests of elephants and humans into conflict; animals this large require vast areas of territory.

A conservationist, Kaddu Sebunya, went to the heart of the problem and summed it up in a few words. She said:

One president said to me: I've never had a voter ask me for more elephants. They want hospitals, education.

Above. Humans have hunted elephants and elephant kinds since time immemorial. This fanciful image of mammoth hunting was painted by Ernest Griset (1844 – 1907) for Sir John Lubbock as part of a series of 19 prehistoric scenes. Reproduced by kind permission of the Lubbock family.

Another, Holly Dublin, addressed the same situation in a rather different way, but it amounts to the same basic idea:
When big banks give out loans, there's a disconnect.
The wildlife comes last.

Above. A shocking collection of bowls and stools made from elephant feet during the nineteenth century. The taking of their ivory was not the only indignity to which elephants were subjected.
Following two pages. Adult African elephants. Photo by David Chancellor.

ELEPHANT 263

The truth is that elephant herds need to be free to go wherever they will in order to find the sustenance they require to live, and this brings them into direct and and open competition with men and women – and populations of men and women are growing at rates that are out of control. Elephants have no regard for the crops and fields so carefully planted by local people; nor do they necessarily show respect for villages and other habitations. A herd is quite capable of trampling and ruining years of work in a very short space of time. Add to this the fact that elephants do not always show the friendly, teddy-bear-like disposition that many people assume is a permanent condition. Adult males in particular can be wild, highly dangerous creatures.

So, in a great many places there is an enormous problem with compatibility between elephants and humans. And usually, with such a divergence of interests, there is only one ultimate winner.

There are various estimates that detail the dramatic fall in elephant populations over the last 50 years or so. The strict accuracy of these estimates is hardly relevant. All are highly alarming and point not only to the incredible suffering of thousands and thousands of individuals, but to the ultimate extinction of elephants within the next century or so.

Many and various are the devices implemented to try to stop this decline. All manner of anti-poaching restrictions are constantly applied – sometimes with considerable degrees of success, sometimes without. But it is very difficult to stop people who live in poverty from harvesting material – in this case ivory – they can sell for amounts that may make a substantial difference to their lives.

The enforcement of bans on ivory poaching go hand in hand, of course, with restrictions on exporting the commodity to those countries where it is wanted. But enforcement of export restrictions runs up against similar problems to those that thwart people who try to prevent the actual poaching. As long as there is a market for illicit products, there will be men and women who find ways around the rules.

Facing page. Wild Asian elephants are under similar threats to those in Africa.
    Photo by Pat Morris.
Pages 266 and 267. A family of African elephants. Photo by Granville Davies.

In Western countries the desire for new ivory is largely dead, and any such desire is regarded as repellent by most people, but in China and other parts of the Far East it is still very much alive. However, ivory poaching (and the desire for new ivory) is something that with time will, hopefully, be curtailed, and indeed China has announced that it will entirely restrict the import of freshly taken ivory.

The other problem – that of mutual incompatibility – runs hand in hand with human population growth and is something that is unlikely to find any easy solution. Yet it would be wrong to suppose that nothing is being done. Large tracts of land have been set aside as animal reserves, and various schemes have been introduced to maintain these and gain the support of local people by offering them means to gain livelihoods (through tourist-related activities, for instance) that are dependent on actually maintaining the animals.

Such developments are going on in many parts of Africa. The government of Botswana has, for instance, been instrumental in securing an ambitious scheme that provides hope for many of its elephants. At the time of this writing, Benin has committed to ensuring the survival of Pendjari National Park, which is described as the largest intact ecosystem in West Africa, for a further 10 years.

Facing page. An African bull elephant staring at the camera. Photo by Granville Davies.

One remarkable development that hopefully allows elephants to live side by side with humans is based on an observation made by Lucy King who works for a number of highly respected conservation-based organizations including Save the Elephants. While working in northern Kenya she noticed that elephants avoided certain trees. On making inquiries with local people concerning this seemingly strange phenomenon, she was told that such trees were home to bees. The people she spoke to went on to tell many stories of elephants becoming highly distressed when enraged bees burst from their hives to sting any encroaching animals, something that the bees did to damaging effect in the area of the trunk, the eyes, and the mouth.

Intrigued by this information, Lucy set about doing some proper practical research into the issue and came up with revealing results. Her work seemed to show that more than 90% of elephants would move rapidly away from any sound of bees within 80 seconds of hearing it.

Having made this discovery, her next problem was to see if the idea could be used to advantage. Could this knowledge be used to stop elephant herds from antagonizing local farmers and rural people with their uncontrollable rampaging through cultivated areas? Her answer was a relatively simple, common-sense one – to create beehive barriers around precious agricultural projects.

Previous two pages. A particularly friendly African elephant. Photo by David Chancellor.
Facing page. A particularly friendly Asian elephant. Photo by Pat Morris.

Facing page and this page. An endangered African elephant has been tranquilized and rendered unconscious so that it can be moved to a safer place by means of a crane, a large vehicle, and a number of helpers. This astonishing photograph was taken by David Chancellor.

This page and facing page. The other side of the coin. This distressed elephant was found appealing for help by photographer David Chancellor and his companions soon after it had been shot by poachers who had rapidly left the scene. The help and comfort given was in vain, however. The animal was so badly injured that it was necessasry to put it out of its pain soon after this photo was taken.

Such barriers could be constructed by creating linked hives hung about 10 metres (33 feet) apart and joined by light pieces of fencing wire. This wire, if disturbed, is likely to cause the bees to emerge in a potentially furious rage and, apparently, elephants emit a low-frequency rumble in response to enraged bees that warns their fellows to keep away.

Not only does this subtle stratagem help in the protection of crops, but it helps preserve water pipes, tanks, grain stores, and property. As a by-product it also allows local people to produce and market honey!

Some other truly useful schemes have been introduced to enable elephants to live in at least some degree of harmony with humans and their inroads into the natural world.

Another remarkable development in Kenya is the building of the Lewa Elephant Underpass. It shows not only the determination of some people to find solutions that are of benefit to elephants, but also the adaptability and intelligence of the elephants themselves.

The project centred around the need for groups of elephants to move from the Ngare Ndare Forest to the forests of Mount Kenya. Among other problems that stood in the way and made safe movement virtually impossible was a very busy major highway that connects the towns of Meru and Nanyuki. A group known as the Mount Kenya Trust proposed that an underpass be constructed in the hope that elephants would learn to use it and not try to cross the road.

Facing page (above). The Lewa Elephant Underpass; *(below)* an elephant using it.

Naturally the proposal came in for criticism and scorn. Many commentators believed that the elephants would never use the tunnel, and it is easy enough to sympathize with such predictions. However, the project went ahead and the underpass was eventually created. Remarkably, those who found the idea flawed were quickly proved wrong. Within days a young elephant that observers called Tony used the new facility, and others soon followed. The underpass became a great success and is used by hundreds of elephants, who all seem to have grasped the entire concept. Curiously, one old male – perhaps in human terms the equivalent of a grumpy old man – refused to submit to the new order. He insisted on running the inevitable risks and stubbornly persisted in crossing the road in his habitual manner.

Although much of this chapter is centred on Africa, similar problems afflict elephants in Asia. The hunting for tusks may not occur on such an enormous scale, but it certainly exists. So too does the conflict over space, and this is not a problem that will go away.

More obvious, however, is the appalling cruelty that is often directed at elephants in captivity. Confined, chained, overworked, viciously beaten, tusks crudely hacked off – captive elephants are subjected to all of these things. Babies are caught in the wild and their mothers often shot before the eyes of the distraught captives. Then they are immediately subjected to a life of torture and drudgery. Many are destined to live their entire lives in conditions that are truly horrific, and there are few signs that such activity will stop.

It is estimated that today there are probably around 40,000 living Asian elephants, and around a quarter of these are in captivity. Only a few centuries ago there were more than 20 times this number living free and wild.

Above. The shadows in this wonderfully evocative photograph by Tim Flack (shown in full on pages 248 and 249) are much larger than the elephants, and more clearly defined. Are we approaching a time when the elephants themselves will be no more than shadows?

Further Reading

There are, of course, many books on elephants and related subjects. This list should not be considered to necessarily represent the finest or most important of them. It is just a very small selection of books that readers may find interesting or informative.

Amranand, P. and Warren, W. (1998). *The Elephant in Thai Life and Legend.*
Anthony, L. (2009, 2017). *The Elephant Whisperer.*
Beard, P. (1963 – and many subsequent editions). *The End of the Game.*
Bloom, S. (2015). *Elephants.*
Bosman, P. and Hail-Martin, A. (1986). *Elephants of Africa.*
O'Connell, C. (2012). *An Elephant's Life.*

Above and below. Day and Night. Asian elephants swimming. Photos by Pat Morris.
Facing page. Asian elephants at sunset. Photo by John Hodges.

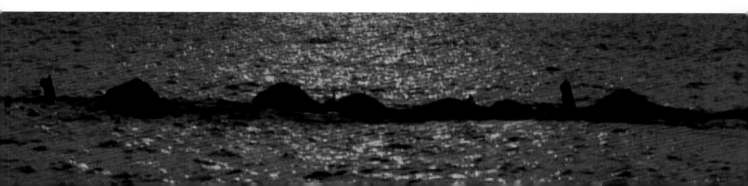

O'Connell, C. (2015). *Elephant Don.*
Pincott, S. (2017). *Elephant Dawn.*
Quammen, D. (1996). *The Song of the Dodo.*
Shoshani, J., ed. (2000). *Elephants: Majestic Creatures of the Wild.*

Sillar, F. and Meyler, R. 1968. *Elephants Ancient and Modern.*
Sukumar, R. (1989). *The Asian Elephant.*
Walker, J. (2009)). *Ivory's Ghosts: The White Gold of History and the Fate of Elephants.*

Index

Alavoine, Jean-Antoine *224-5*
Ali Muhammed 185
Al-Thani, Sheikh Saud *128, 129*
Amal-e Hashim *227,* 228
Amber, India 100
Amebelodon 114, *118*
American Museum of Natural History 176
Angkor Wat 165
Annick at Aldo Workshop *Title page,* 6, *32-3, 174-5*
Anthony, Lawrence 187-9, 191
Apes 146
Apocrypha, The 201
Arcimboldo, Giuseppi 216, *217*
Aristotle 8
Atlas Mountains 76
Austin, Denise 150, *150*
Austin, Jill 6

Babar, The Elephant 208, *210*
Baines, T. *192-3,* 228, *246, 247*
Bali 58
Bangkok 222
Barclay, T. *Half title,* 6, *80*
Barnum, Phineas 185, *186*
Barye, A. 229
Barytherium 118
Bastille, Elephant, the *224-5*
Beard, Peter 6, *55, 67, 140,* 207, *207,* 233, 252-2

Belfast Zoo 149
Bengal 180
Benin 248
Berezovka Mammoth 134, *135*
Bering Strait 126, 137
Bosch, Hieronymous *215,* 216
Botswana 268
Bowring, John 106
Breuer, Thomas 6, *75*
Brunhoff, Jean 208
Bryce Canyon 58
Buddha 196-7
Buffalo, Cape 50
Bugatti, Rembrandt 229
Burian, Zdenek *115, 124-5, 130*

Caesar, Julius *212,* 213
California 119
Cambodia 165
Carthage 76
Chancellor, David 6, *46, 51, 68-9, 164, 236-7, 262-3, 274-5, 276-7*
China 268
Ching, Raymond 6, 26-7, *166-7,* 233, 238
Chuckchi Sea 137
Chunee,The Elephant 180-1,*181*
Clarkson, Glyn 6, *108-9*
Coney Island 182, *182*
Congo *75,* 76, *162, 163*

Cross, Edward 180
Cruikshank, George *181*
Cyprus 118

Dali, Salvador 233
Davies, Granville 6, *38, 48-9, 59, 250, 253, 258-9, 266-7, 269*
Deinotherium 115 118
Denis, Armand 173, 176
Dierenpark, Emmen *190,* 191
Dima 134, *135*
Disney, Walt 208
Dodo 58, 119
Dolphins 146
Dublin, Holly 261
Dugongs 112
Dumbo 208
Durer, Albrecht 228

Egypt 161
Eleazor 201, *202,* 203
Elephantographia Curiosa 204, *204, 205, 206*
Elephants
  African 44-77, 81, 161, 228
  Appearance 35-40
  Asian 78-107, 161, 195-7, 201
  Behaviour 31-5, 47, 61, 64-6, 70-1, 96
  Captivity 81, 86, 93, 96, 144-6, 149, 150, *150, 151,* 151, 152, *152*

Page 286. A small African elephant says goodbye. Photo by Pat Morris.

153, 158, 161-2, *163, 164, 169, 170,* 171,*172,* 173, 180, *181, 182,* 182-3, *184,* 185-6. *186,* 213, 218, *219,* 229
Circus 153, *154, 240, 243, 244*
Conservation 248-281
Destructiveness 54, 58-9, 263
Evolution 111-2, 114
Forest elephants 74, *75,* 76
Graveyards 141
Intelligence 144-6, 149, 155, 158, 189, 191, 220, *220,* 221, *222,* 223
Painting *138-9,* 144, 145 *145,* 146
Prehistoric 108-137
Pygmy 104, *105,* 106
Statistics 51, 54
White 168, *169,* 170
Elephas maximus 106
Eocene 114, 118
Etosha, Namibia 6, *10-13, 44-5, 56-7, 77* 208
Ewaso, Nyiro River *68-9*
Exeter Exchange 180, *181*

Flack, Tim 6, *7, 72-3, 248-9, 281*
Florence 219
Ford, Walton 6, 233, *234-5,* 236, 238
Frazier, Joe 185

Gabon 81, 162
Ganesh *200,* 201, *220*
Gongala-na-Bodio 162
Geographical Society 195
Giza 137
Graphic, The 225

Griset, Ernest 260
Groening, G. *226,* 228
Guinea Pigs 111

Hampton Court 212
Hannibal 76, *160,* 161, 201, 213
Hansken, the Elephant *218,* 218-219
Hartenfels, Georg 204
Hippopotamus 50, 111
Hodges, John *Endpapers, 5,* 6, *14-5, 28, 34, 36-7, 43, 84, 85, 88-9, 107, 113, 233*
Hohenwold, Tennesse 153
Holiday, G. *254*
Horley, Surrey 151, 152, *152*
Hyrax 111, 112, *112*

Ivory 40, 42, *60,* 61, *254,* 255-6, 257, 264, 268

Jaipur *80, 86*
Jamin, Paul joseph *123*
Jardin des Plantes 163
Jesus *221*
Jumbo, the Elephant 180, *183, 184,* 185, 186, *187*
Jungle Book, The 208, *209*

Kenya *68-9,* 251, 272, 278
Khajuraho *196, 197*
Kilimanjaro 61
King, Lucy 272
Kipling, Rudyard 208, *209*
Knight, Charles R. *127*
Knight, Hilary 6
Kochabua, Prateep, 6, 238-9
Kuhnert, Wilhelm 229, *232*

La Baume Latrone *132,* 134
Laplume, Officer 162
Leo X 172, 173
Leopold II 81, 162
Le Petit Journal 170-1
Lewa 278, *279*
Livingstone, David 228,
Lizzie, the Elephant 150, *151*
London Zoo 133 151, *151,* 228, *229,* 241
Loubiere, Simon de la 158
Loxodonta africana 47
Loxodonta cyclotis 74 , *75*
Lubbock, Lyulph 6, *260*

Macklin Bible 201, *202*
Maclean, Diana 6
Madison Square Garden 185
Mahabalipuram 8, *18-9*
Malta 113
Mammoths *110,* 119, *120, 121, 123, 124-5, 126, 128, 128, 129,* 213
Manatees 112, *112*
Mantegna Andrea *212,* 213, 214
Manuel I 173
Ma-Robert and the Elephant 192-3, 228, 247
Mastodons 119, *122,* 126, *127*
Mauritius 119
Mboli River *75*
Metcalf, John 6, *105*
Miocene 114, 137
Morris, Pat 6, *16-7, 39, 82-3, 87, 90-1, 92, 101, 102-3, 142-3, 147, 148, 156-7, 265, 273, 286*
Mumfie 208, *211*
Musth 70, 71

Namibia 6, *10-13, 44-5, 56-7, 77,* 208

286 INDEX

Osborn, Henry 112, 114

Palmer, Irene 6, *35, 41, 44-5, 52-3, 77, 100, 56-7, 196-7*
Paris *224-5*
Peerless, Bury 6, *18-9, 20-1,* 30, *78-9,* 86
Pendjari National Park 268
Petrou, Peter 6, 200
Platybelodon 114, *116-7*
Pliny, the Elder 25
Prado, The 214, *215*
Prasarttong-Osoth, Dr. Prasert 5, 6, *110,* 176, *177*
Proboscidea 114
Punic Wars 76

Quammen, David 58

Rangoon 152
Raphael 172, 173
Rembrandt 218, *219,* 228
Rhinoceros 111
Ribeiro, Jose 160
Ringling Brothers *244*
Ripenda, Jacopo 160
Ritchie, James 153

Roman Empire 76
Romano, G. *172,* 173
Ropes, Chelsea *64*
Rouffignac Cave *132,* 134
Rubens, Peter Paul *214,* 218

Scharf, G. *227,* 228
Scientific American 152, 153
Sebunya, Kaddu 260
Sheffield 151
Sheila, the Elephant 149, 150, *150*
Shepherd, David 233
Siam 31, 106, 168, 173
Sirijirasuk, Rattapol 6, *94-5, 97, 98-9, 110,* 114, *116-7, 120-1, 131, 136,* 137, *138-9, 145, 165, 169, 178-9*
Smith, Jenny *50*
Song of the Dodo 58
Sri Lanka 8, *9*
Steller's Sea Cow 112
St. Petersburg 134, *135*
Sudan 183
Sukhothai *22-3, 104, 110,* 115, *120-1, 165,* 176, *177, 195,* 200, 222, *223*

Tervuren *174-5,* 176
Thailand *104, 110,* 115, *115, 120-1, 131, 136, 137, 138-9, 145,* 158, 176, *177, 178-9,* 195, 200, *223*
Thula Thula Game Reserve 6, *50, 64,* 168, *188,* 189
Tom Thumb, the Elephant 186
Topsy, the Elephant 180, *182,* 182
Tozer, Katherine 208

Vachajitpan, Piriya 6
Victoria, Queen 185

White Elephant 168, *169,* 172, 173
Wolf, Joseph 229
World of Elephants Museum *104, 110, 115,* 115, *120-1,* 129, *131, 136, 137,* 165, 176, *177, 178-9,* 200, *223*
Wrangel Island 137

Zambezi River 228

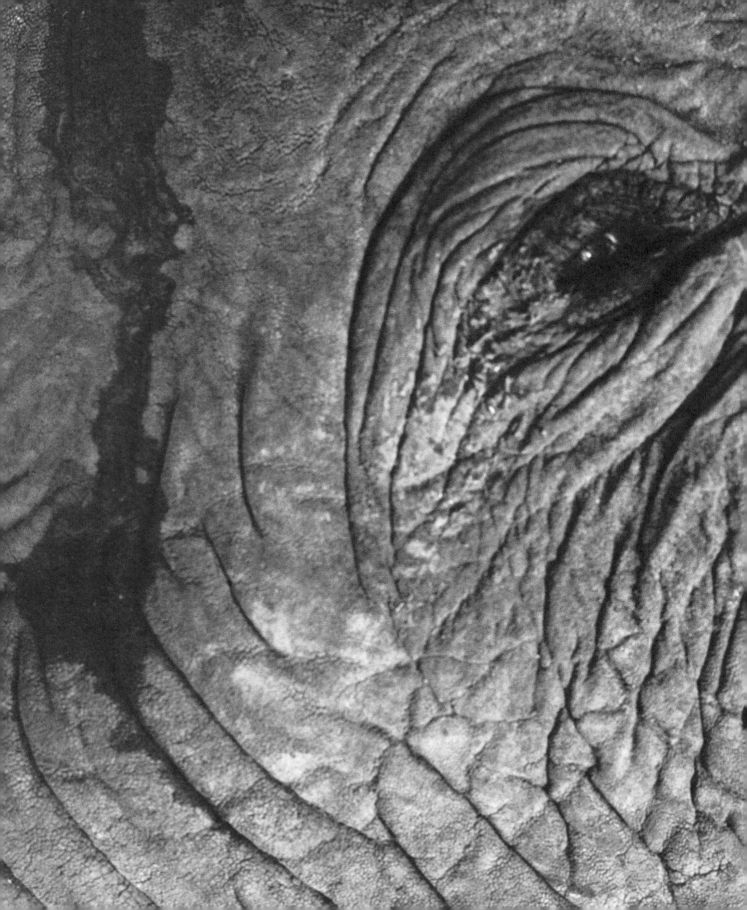